Behaviour of Electromagnetic Waves in Different Media and Structures

Editor

Fredrik Jonasen

scitus academics

Behaviour of Electromagnetic Waves in Different Media and Structures
Edited by **Fredrik Jonasen**

ISBN: 978-1-68117-281-1
Library of Congress Control Number: 2016937453

SCITUS Academics LLC,
www.scitusacademics.com
Box No. 4766, 616 Corporate Way,
Suite 2, Valley Cottage,
NY 10989

Notice

Printed in United States of America on Acid Free Paper

Preface

The term electromagnetic wave describes the electromagnetic radiation (EMR) movement through space. Different forms of EMR differ in their wavelengths, which vary from many yards to a distance smaller than the diameter of the atomic nucleus. The entire range goes from radio waves to microwaves, visible light, ultraviolet and X-rays to gamma rays and is known as the electromagnetic spectrum in decreasing wavelength range. Electromagnetic waves are formed when an electric field matches a magnetic field. The magnetic and electric fields of an electromagnetic wave are perpendicular to each other and the direction of the wave. In science and everyday life, electromagnetic waves are used. Radio waves, TV waves and microwaves are all types of electromagnetic waves. They differ from each other in wavelength. Waves in the electromagnetic spectrum differ in size from very long radio waves to very short gamma-rays. This book emphasizes on the behavior of wave propagation and electromagnetic wave computing in different complex media. The chapters contain authoritative and complicated analytical and numerical solutions to their electromagnetic problems in complex media and geometries. This will be of immense valuable book for engineers and practitioners in electromagnetics and microwave.

Table of Contents

Chapter 1

Electromagnetic Waves in Cavity Design

Suk Kim Hyoung[1]

[1] *Kyungpook National University, Korea*

1. INTRODUCTION

Understanding electromagnetic wave phenomena is very important to be able to design RF cavities such as for atmospheric microwave plasma torch, microwave vacuum oscillator/amplifier, and charged-particle accelerator. This chapter deals with some electromagnetic wave equations to show applications to develop the analytic design formula for the cavity. For the initial and crude design parameter, equivalent circuit approximation of radial line cavity has been used. The properties of resonator, resonant frequency, quality factor, and the parallel-electrodes gap distance have been considered as design parameters. The rectangular cavity is introduced for atmospheric microwave plasma torch as a rectangular example, which has uniform electromagnetic wave distribution to produce wide area plasma in atmospheric pressure environment. The annular cavity for klystrode is introduced for a microwave vacuum oscillator as a circular example, which adapted the grid structure and the electron beam as an annular shape which gives high efficiency compared with conventional klystrode. Some simulation result using the commercial software such as HFSS and MAGIC is also introduced for the comparison with the analytical results.

2. EQUIVALENT CIRCUIT APPROXIMATION OF RADIAL-LINE CAVITY

Microwave circuits are built of resonators connected by waveguides and coaxial lines rather than of coils and condensers. Radiation losses are eliminated by the use of such closed elements and ohmic loss is reduced because of the large surface areas that are provided for the surface currents. Radio-frequency energy is stored in the resonator fields. The linear dimensions of the usual resonator are of the order of magnitude of the free-space wavelength corresponding to the frequency of excitation. A simple cavity completely enclosed by metallic walls can oscillate in any one of an infinite number of field configurations. The free oscillations are characterized by an infinite number of resonant frequencies corresponding to specific field patterns of modes of oscillation. Among these frequencies there is a smallest one,

$$f_0 = c/\lambda_0 \tag{1}$$

, where the free-space wavelength is of the order of magnitude of the linear dimensions of the cavity, and the field pattern is unusually simple; for instance, there are no internal nodes in the electric field and only one surface node in the magnetic field.

The oscillations of such a cavity are damped by energy lost to the walls in the form of heat. This heat comes from the currents circulating in the walls and is due to the finite conductivity of the metal of the walls. The total energy of the oscillations is the integral over the volume of the cavity of the energy density,

$$W = \frac{1}{2}\int_v \left(\varepsilon_0 E^2 + \mu_0 H^2\right) dv \tag{2}$$

$$\mu_0 = 4\pi \times 10^{-7}\, H/m \text{ and } \varepsilon_0 = \frac{1}{36\pi} \times 10^{-9}\, F/m \tag{3}$$

, where E and H are the electric and magnetic field vectors, in volts/meter and ampere-turns/meter, respectively. The cavity has been assumed to be empty. The total energy W in a particular mode decreases exponentially in time according to the expression,

$$W = W_0 e^{-\frac{\omega_0 t}{Q}} \tag{4}$$

, where $\omega_0 = 2\pi f_0$ and Q is a quality factor of the mode which is defined by

$$Q = \frac{2\pi(\textit{energy stored in the cavity})}{(\textit{energy lost in one cycle})}. \tag{5}$$

The fields and currents decrease in time with the factor $e^{-\frac{\omega_0 t}{Q}}$. Most klystrons and klystrodes are built with cavities of radial-line types. Several types of reentrant cavities are shown in Fig. 1. It is possible to give for the type of cylindrical reentrant cavities a crude but instructive mathematical description in terms of approximate solutions of Maxwell's equations.

The principle, or fundamental, mode of oscillation of such cavity, and the one with the longest free-space wavelength, has electric and magnetic fields that do not depend on the azimuthal angle defining the half plane though both the axis and the point at which the fields are being considered. In addition, the electric field is zero only at wall farthest apart from the gap and the magnetic field is zero only at the gap. In this mode the magnetic field is everywhere perpendicular to the plane passing through the axis the electric field lies in that plane. Lines of magnetic flux form circles about the axis and lines of electric flux pass from the inner to the outer surfaces.

In the principle mode of radial-line cavity only Ez, and Hz are different from zero and these quantities are independent of ϕ. (see Fig. 2 for cylindrical coordinates and dimensions of the cavity). The magnetic field automatically satisfies the conditions of having no normal component at the walls.

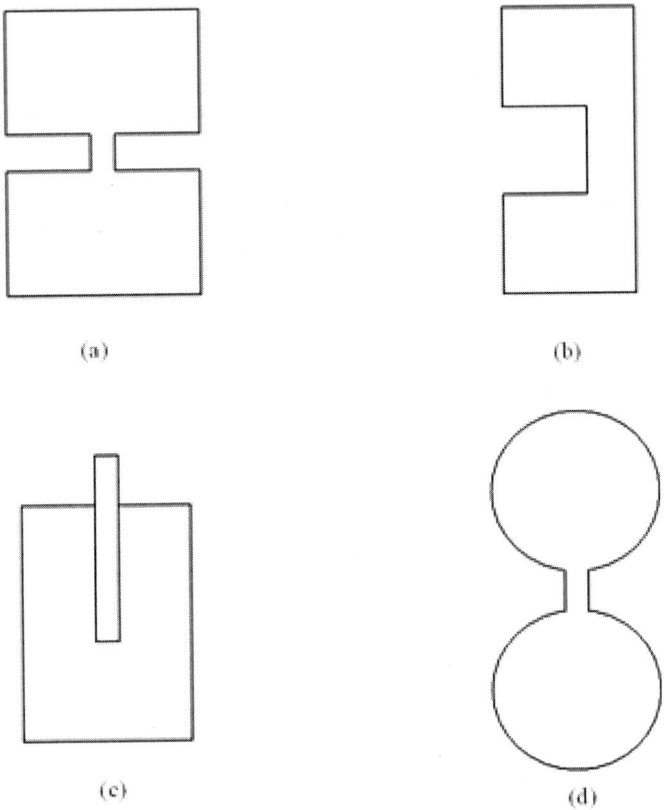

Figure 1. Resonant cavities; (a) Coaxial cavity, (b) Radial cavity, (c) Tunable cavity, (d) Toroidal cavity

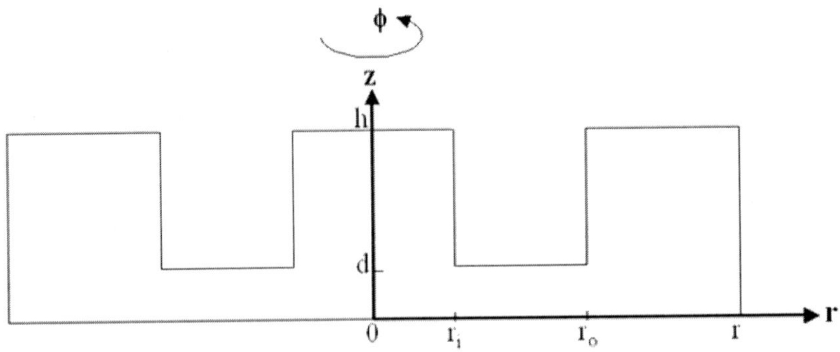

Figure 2. Cylindrical coordinates and dimensions of the radial cavity

The cavities in which RF interaction phenomena happens with charged particles almost always have a narrow gap, that is, the depth of the gap *d* (see Fig. 2), is small compared with the radius r_0 of the post (*d((r0* in Fig. 2).. If the radius of the post is much less than one-quarter of the wavelength, and if the rest of the cavity is not small, the electric field in the gap is relatively strong and approximately uniform over the gap. It is directed parallel to the axis and falls off only slightly as the edge of the gap is approached. On the other hand, the magnetic field increases from zero at the center of the gap in such a manner that it is nearly linear with the radius.

In a radial-line cavity the electric field outside the gap tends to remain parallel to the axis, aside from some distortion of the field that is caused by fringing near the gap; it is weaker than in the gap and tends to become zero as the outer circular wall is approached. The magnetic field, on the other hand, increases from its value at the edge of the gap and has its maximum value at the outer circular wall.

It is seen that, whereas the gap is a region of very large electric field and small magnetic field, the reentrant portion of the cavity is a region of large magnetic field and small electric field. The gap is the capacitive region of the circuit, and the reenetrant portion is the inductive region. Charge flows from the inner to the outer conducting surface of the gap by passing along the inner wall, across the outer end. The current links the magnetic flux and the magnetic flux links the current, as required by the laws of Faraday, Biot and Savart.

2.1. Capacitance in cylindrical cavity

If the gap is narrow, the electric field in the gap is practically space-constant. Thus the electric field *Ez* in the gap of the circular cavity (see Fig. 2) comes from Gauss' law,

$$E_z = \frac{\sigma}{\varepsilon_0} = \frac{Q}{\varepsilon_0 \pi (r_0^2 - r_i^2)}. \tag{6}$$

At the end of the cavity near to the gap both Ez and Er exist and the field equations are more complicated. If d is small compared with h and r, it can be assumed that the fields in the gap are given approximately by the preceding equation.

Therefore,

$$C = \frac{Q}{V} = \frac{Q}{E_z d} = \frac{\varepsilon_0 \pi (r_0^2 - r_i^2)}{d} \quad (7)$$

2.2. Inductance in cylindrical cavity

The magnetic field $H\varphi$ comes from Ampere's law,

$$\oint H r d\theta = I \quad (8)$$

Therefore,

$$B = \frac{\mu_0 I}{2\pi r} \quad (9)$$

The total magnetic flux in the cylindrical cavity,

$$\Phi = \int_{r_0}^{r} Bh dr = \frac{\mu_0 I h}{2\pi} \int_{r_0}^{r} \frac{dr}{r} = \frac{\mu_0 I h}{2\pi} \ln \frac{r}{r_0} \quad (10)$$

Comparing this with inductance definition, $\Phi = LI$, we get the followings;

$$L = \frac{\mu_0}{2\pi} h \ln \frac{r}{r_0} \quad (11)$$

2.3. Resonance frequency in cylindrical cavity

The resonant wavelength of a particular mode is found from a proper solution of Maxwell's equation, that is, one that satisfies the boundary conditions imposed by the cavity. When the walls of the cavity conduct perfectly, these conditions are that the electric field must be perpendicular to the walls and the magnetic field parallel to the walls over the entire surface, where these fields are not zero.

The resonant frequency $f0$ could be calculated for the principal mode of the simple reentrant cavity. The resonant cavity is modeled by parallel LC circuits as can be seen in Fig. 3. In fact, cavities are modeled as parallel resonant LC circuits in order to facilitate discussions or analyses. The resonant frequency is inversely proportional to the square root of inductance and capacitance;

$$\omega = \frac{1}{\sqrt{LC}} = \frac{1}{\sqrt{\frac{\varepsilon_0 \mu_0}{2} \frac{r_0^2 - r_i^2}{d} h \ln \frac{r}{r_0}}} . \qquad (12)$$

2.4. Unloaded Q in cylindrical cavity

In the cavity undergoing free oscillations, the fields and surface currents all vary linearly with the degree of excitation, that is, a change in one quantity is accompanied by a proportional change in the others. The stored energy and the energy losses to the walls vary quadratically with the degree of excitation.

Since the quality factor Q of the resonator is the ratio of the stored energy and the energy losses per cycle to the walls, it is independent of the degree of excitation.

$$Q = \frac{2\pi f(\text{energy stored in the cavity})}{(\text{power lost})} = \omega_0 \frac{U}{P_{loss}} = \omega \frac{\frac{1}{2} LI^2}{\frac{1}{2} RI^2} = \omega \frac{L}{R} . \qquad (13)$$

The resonator losses per second, besides being proportional to the degree of excitation, are inversely proportional to the product of the effective depth of penetration of the fields and currents into the walls, the skin depth, and the conductivity of the metal of the walls. Since the skin depth is itself inversely proportional to the square root of the conductivity, the losses are inversely proportional to the square root of the conductivity. The losses are also roughly proportional to the total internal surface area of the cavity; and this area is proportional to the square of the resonant wavelength for geometrically similar resonators. The skin depth is proportional to the square root of the wavelength, and hence the losses per second are proportional to the three-halves power of the resonant wavelength. The loss per cycle, which is the quantity that enters in Q, is proportional to the five-halves power of the resonant wavelength. Since the energy stored is roughly proportional to the volume, or the cube of the wavelength, the Q varies as the square root of the wavelength for geometrically similar cavities, a relationship that is exact if the mode is unchanged because the field patterns are the same. In general, large cavities, which have large resonant wavelengths in the principal mode, have large values of Q. Cavities that have a surface area that is unusually high in proportion to the volume, such as reentrant cavities, have Q's that are lower than those of cavities having a simpler geometry.

The surface current, J, is equal in magnitude to $H\phi$ at the wall. The power lost is the surface integral over the interior walls of the cavity.

$$P_{loss} = \frac{R_s}{2}\int J^2 ds = \frac{R_s}{2}\int |H_\phi|^2 ds$$

$$= \frac{R_s}{2}\left[\left(\frac{I}{2\pi r_o}\right)^2 2\pi r_0(h-d) + 2\int_{r0}^{r}\left(\frac{I}{2\pi r_0}\right)^2 2\pi r dr + \left(\frac{I}{2\pi r_o}\right)^2 2\pi r h\right]$$

$$= \frac{R_s I^2}{2\pi}\left[\frac{h-d}{r_0} + 2\ln\frac{r}{r_0} + \frac{h}{r}\right] \equiv \frac{1}{2}RI^2, \tag{14}$$

where the shunt resistance is

$$R \equiv \frac{R_s}{2\pi}\left[\frac{h-d}{r_0} + 2\ln\frac{r}{r_0} + \frac{h}{r}\right]. \tag{15}$$

The surface current can be considered concentrated in a layer of resistive material of thickness. Surface resistance is that

$$R_s = \frac{1}{\sigma\delta} = \frac{1}{\sigma\sqrt{\frac{1}{\pi f \mu \sigma}}} = \sqrt{\frac{\pi f \mu}{\sigma}}. \tag{16}$$

As an example, the conductivity of copper is $\sigma_{copper} = 5.8\times 10^7 \, / \, \Omega m$ and since copper is nonmagnetic

$\mu_{copper} = \mu_0 = 4\pi \times 10^{-7} H / m$, hence, in case of that cavity material is copper, for f=6GHz,

$$\delta = 0.85\mu m$$
$$R_s = 2.02 \times 10^{-2} \Omega$$
$$R = 2.66 \times 10^{-3} \Omega$$
$$L = 1.01 \times 10^{-10} H$$
$$Q_0 = \omega \frac{L}{R} = 1431. \tag{17}$$

The shunt conductance G is, as given by the expression,

$$G = \frac{(energy\ lost\ per\ second)}{\overline{V(t)^2}} \tag{18}$$

is defined only when the voltage $V(t)$ is specified. In a reentrant cavity the potential across the gap varies only slightly over the gap if the gap is narrow and the rest of the cavity is not small. A unique definition is obtained for G by using for $V(t)$ the potential across the center of the gap. The gap voltage is proportional to the degree of excitation, and hence the shunt conductance is independent of the degree of excitation. For geometrically similar cavities the shunt conductance varies inversely as the square root of the resonant wavelength for the same mode of excitation. This relationship exists because for the same excitation $\overline{V(t)^2}$ is proportional to the square of the wavelength and the loss per second to the three-halves power of the wavelength.

2.5. Lumped-constant circuit representation

The main value of the analogy between resonators and lumped-constant circuits lies not in the extension of characteristic parameters to other geometries, in which the analogy is not very reliable, but in the fact that the equations for the forced excitation of resonators and lumped-constant circuits are of the same general form.

If, for example, it is assumed that the current i(t) passes into the shunt combination of L, C and conductance G, by Kirchhoff's laws, (see Fig. 3)

$$i(t)=C\frac{dV(t)}{dt}+\frac{1}{L}\int V(t)dt+GV(t). \tag{19}$$

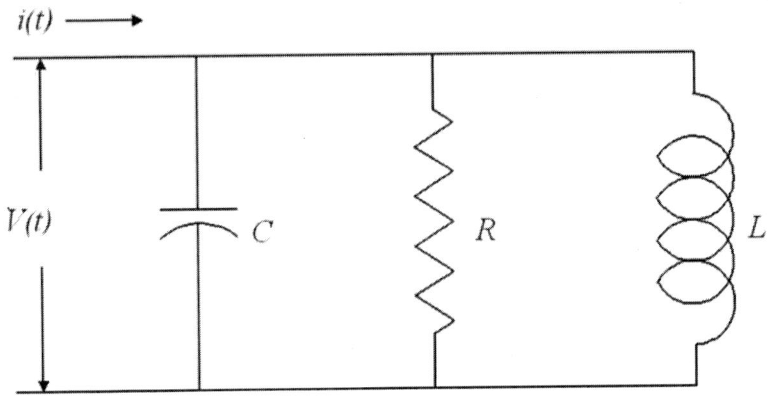

Figure 3. Limped-constant circuit

On taking the derivative and eliminating L,

$$\frac{di(t)}{dt}=C\left[\frac{d^2V(t)}{dt^2}+\omega_0^2V(t)\right]+G\frac{V(t)}{dt}. \tag{20}$$

In other word,

$$\frac{\omega}{C}\frac{di(t)}{d\theta}=\omega^2\frac{d^2V(t)}{d\theta^2}+2\gamma\omega\frac{dV(t)}{d\theta}+\omega_0^2V(t), \tag{21}$$

where $\gamma=G/2C$, $\omega_0=1/\sqrt{LC}$ and $\vartheta=\omega t$, which are used to calculate numerically the initial beam effect in the last chapter.

For a forced oscillation with the frequency ω,

$$i_\omega=\left[G+j\omega_0C\left(\frac{\omega}{\omega_0}-\frac{\omega_0}{\omega}\right)\right]V_\omega. \tag{22}$$

Thus, there is defined circuit admittance

$$Y = G + j\omega_0 C\left(\frac{\omega}{\omega_0} - \frac{\omega_0}{\omega}\right). \quad (23)$$

These equations describe the excitation of the lumped-constant circuit.

3. NUMERICAL ANALYSIS FOR THE HIGH FREQUENCY OSCILLATOR SYSTEM WITH CYLINDRICAL CAVITY

In this section, we will meet an circular cavity example of a klystrode as a high frequency oscillator system with the knowledge which is described in previous sections.

Conventional klystrodes and klystrons often have toroidal resonators, i.e., reentrant cavity with a loop or rod output coupler for power extraction. These resonators commonly use solid-electron-beam which could limit the output power. One way to get away this limitation is to use the annular beam as was commonly done in TWTs. The main reason using reentrant cavities in most microwave tubes with circular cross sections is that the gap region should produce high electric field and thus high interaction impedance of the electron beam when the cavity is excited. In our design we assume a short cavity length, *d*, along the longitudinal direction parallel to the electron motion. In the meantime the width of electron beam tunnel, *r*0−*ri*, is much larger, i.e. *d*((*r*0−*ri* as shown in Fig 4. And thus the efficiency of beam and RF interaction in this klystrode cavity depends sensitively upon the cavity shape at the beam entrance of the RF cavity in the beam tunnel. A simple trade-off study suggests to put to use of gridded plane, so-called a cavity grid (anode), so that the eigenmode of the reentrant cavity is maintained. With the gridded plane removed and left open, the TM01-mode has many competing modes and the interaction efficiency disappears. The use of thin cavity grid in the beam tunnel, however, can slightly reduce the electron beam transmission, which will not pose a much of problem when the same type of grid is used in between the cathode and anodic cavity grid. In the simulations with the MAGIC and HFSS codes, the anodic cavity grid could be assumed to be a smooth conducting surface, and pre-bunched electrons were launched from those surfaces of cavity grid. This kind of concept can provide a compact microwave source of low cost and high efficiency that is of strong interest for industrial, home electronics and communications applications.

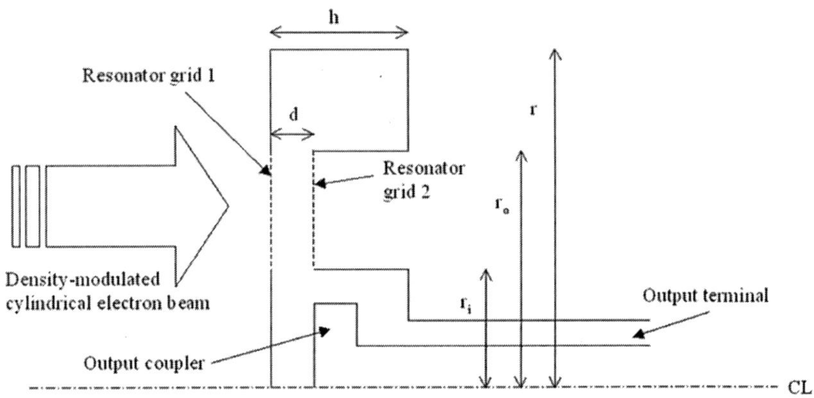

Figure 4. Schematics of the annular beam klystrode with the resonator grids for the high electric field and high interaction efficiency in the gap region. This cavity structure allows easier power extraction through the center coax coupler

The klystrodes consist of the gated triode electron gun, the resonator and the collector. The gated electron gun provides with the pre-modulated electron bunches at the fundamental frequency of the input resonator, where the voltage on the grid electrode is controlled by an external oscillator or feedback system. The other possible type of gated electron guns could be the field-emitter-array gun, RF gun, and photocathode. The electron bunches arrive at the output gap with constant kinetic energy but with the density pre-modulated. Here, we assumed the electron beam is operated on class B operation, that is, electron bunch length is equal to one half of the RF period. Through the interaction between electron beam and RF field, the kinetic energy is extracted from the pre-modulated electrons and converted into RF energy.

Figure 4 shows the schematics of the circular gridded resonator with center coupling mechanism for the easy and efficient power extraction. In this section, we will describe the design of annular beam klystrode in C-band.

3.1. RF interaction cavity design

As we have seen in previous section, using the lumped-circuit approach, the resonant frequency of this protuberance cavity with the annular beam is expressed as

$$\omega = \frac{1}{LC} = \frac{1}{\sqrt{\frac{\varepsilon_0 \mu_0}{2} \frac{r_0^2 - r_i^2}{d} h \ln \frac{r}{r_0}}} \quad (24)$$

Since this expression is an approximation which gives the tendency of frequency variation when we are adjusting design parameters, we can perform parameter tuning exercise using design tools such as HFSS. Fig. 5 shows an example of the detailed design using HFSS where the emission was introduced at the gap region between inner radial distances of 5.7 and 9.4mm. In the figure, the electric field is enhanced and fairly uniform due to the presence of resonator grid1 and resonator grid2. The grid structure in beam inlet and beam outlet make the electric field maintain fairly high intensity in the gap region through which the electron beam passes to interact with RF. Figure 6 also show scattering parameter plots where resonator grids of the klystrode are considered closed metal wall and the cavity has only output terminal as one port system. The bold line is the real value of S and the thin line is the imaginary one. The resonator frequency is 5.78 GHz in the absence of finite conductivity of cavity and electron beam.

The detailed tuning of beam parameters for efficient klystrode could be investigated using PIC code such as MAGIC. As an example, the current is assumed density-modulated in the input cavity and cut-off sinusoidal,

$$I(z=0,t) = I_0 MAX\left(\sin\left(\omega t\right),0\right)$$
$$= I_0\left[\frac{1}{\pi}+\frac{1}{2}\sin\left(\omega t\right)-\sum_{n-0}^{\infty}\frac{2}{\pi(4n^2-1)}\cos\left(2n\omega t\right)\right] \quad (25)$$

whose peak current, I_{peak}, is 3 amperes.

The tube is supposed of being operated in class B as shown in Fig. 7. A class B amplifier is one in which the grid bias is approximately equal to the cut-off value of the tube, so that the plate current is approximately zero when no exciting grid potential is applied, and such that plate current flows for approximately one-half of each cycle when an AC grid voltage is applied. *I*=*I0MAX*(sin(*ωt*),0)

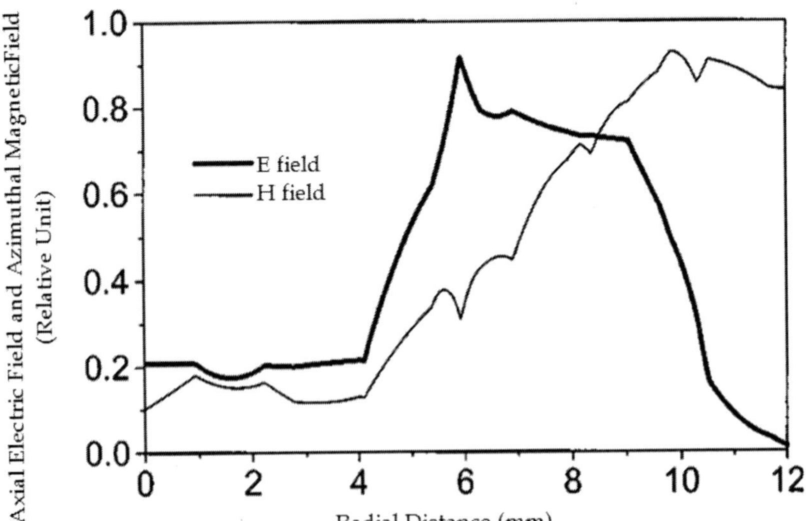

Figure 5. Magnitude of axial electric field and azimuthal magnetic field (in relative unit) along the radial distance on the mid-plane between resonator grid 1 and grid 2 in the cavity. Emission surface is between the radial distances of 5.7 and 9.4 mm

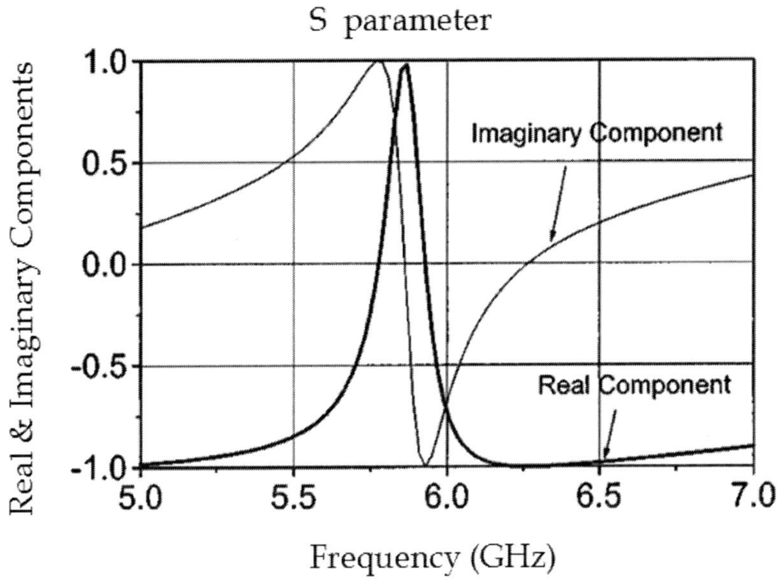

Figure 6. Scattering parameter plots. The resonator frequency is 5.78 GHz in the absence of finite conductivity of cavity and electron beam

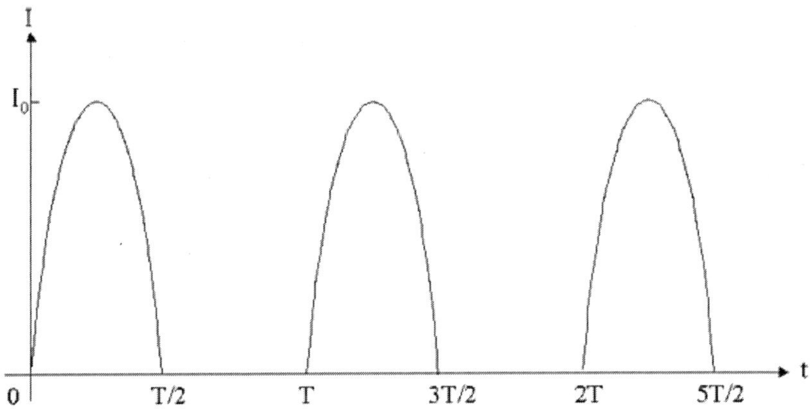

Figure 7. Pre-modulated electron beam in current vs. time; cut-off sinusoidal current which is used in class B operation,

The fundamental mode (TM_{01}-mode) to be interacted with longitudinal traversing electron beam was adapted to our annular beam resonators for the high efficiency device.

Electron transit angle between electrodes gives limitation in the application of the conventional tubes at microwave frequencies. The electron transit angle is defined as

$$\beta = \omega\tau_g = \omega d/v_0 , \tag{26}$$

where $\tau_g = d/v_0$ is the transit time across the gap, d is separation between cathode and grid, $v_0 = \sqrt{2eV_0/m} = 0.593\times10^6\sqrt{V_0}$ is the velocity of the electron, and V0 is DC voltage.

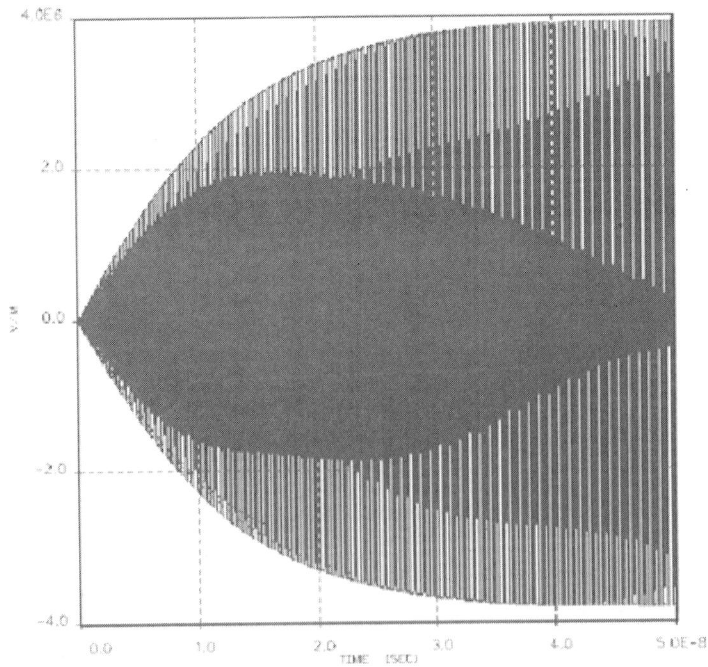

Figure 8. Electric field in the gap region across the anode electrode 1(grid1) and electrode 2(grid2); The field reaches 4,000,000 V/m

The transit angle was chosen to give that the transit time is much smaller than the period of oscillation for the efficient interaction between RF and electron beam, so that, the beam coupling coefficient, *M*, is 0.987. The resonant frequency is 5.78 GHz in cold cavity and 6.0 GHz in hot cavity. Although the frequency shift may be greater than the value of normal case, this would be come from the fact that this annular beam covers much more area with electron beam than the conventional solid beam in a given geometry. As we can see in Fig. 8 and Fig. 9, this resonant cavity is filled and saturated with the RF power in 50 ns, and reveals high efficiency of about 67%. The output power is 1250 W so that the efficiency of this annular beam klystrode reveals 67 % at 6.004 GHz.

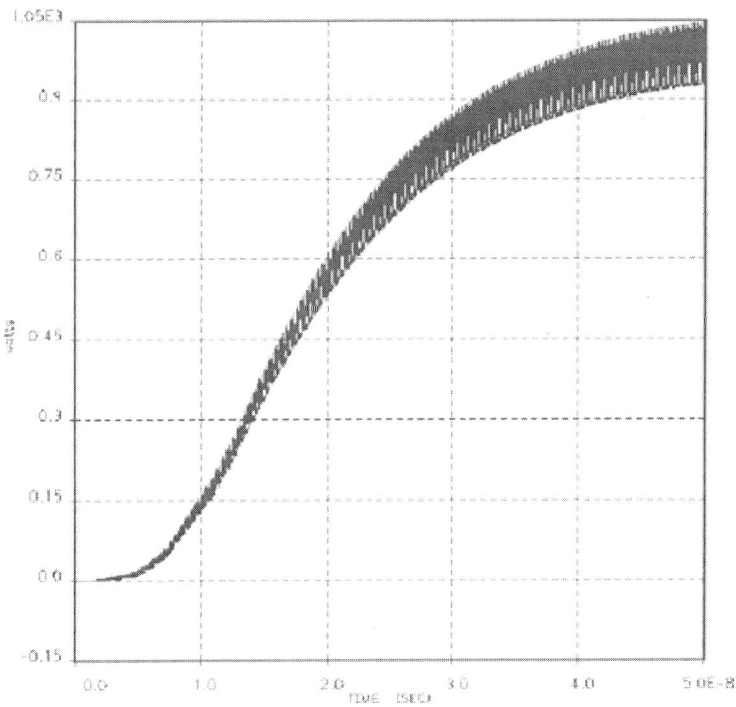

Figure 9. Output power going through the output port vs. time where driving frequency is 6GHz. It goes to about 1.25kW

3.2. RF interaction efficiency calculation

There are some computational design codes for the klystrode. But in this section, 1-dimensional but realistic electron beam and electric field shape are introduced to develop analytical calculations for the klystrode design, which results in easy formulas for the efficiency and electric field in the gap region of the klystrode in steady state.

Maxwell's equations for electron beams are followings,

$$\nabla \cdot E = -\frac{\rho}{\varepsilon}, \tag{27}$$

$$\nabla \times E = -\frac{\partial B}{\partial t} \tag{28}$$

and

$$\nabla \times H = J + \frac{\partial D}{\partial t}. \tag{29}$$

As an approximation, the electron dynamics is in 1-D space. Then, the Maxwell's equations are simplified as the followings.

$$\frac{\partial E}{\partial z} = -\frac{\rho}{\varepsilon}, \tag{30}$$

$$\frac{\partial B}{\partial t} = 0 \tag{31}$$

and

$$-\rho v + \varepsilon \frac{\partial B}{\partial t} = \varepsilon \frac{\partial E}{\partial z} v + \varepsilon \frac{\partial E}{\partial t} = \varepsilon \frac{dE}{dt} = 0. \tag{32}$$

Therefore, this means that E remains constant for each electrons moving with velocity, v.

From the Lorentz force equation,

$$\frac{dv}{dt} = -\frac{e}{m}E = (const.) \tag{33}$$

for each particle with velocity v Define the snapshot time be τ such that,

$$\tau = t_x + t_y, \tag{34}$$

where tx is transit time for the moving particle from resonator grid1 to the transit distance, z

, and ty is leaving time for moving particle from the resonator grid1. Its definition is shown in schematic representation for the transit time, departure time, snapshot time, and transit distance in Fig. 10. Therefore, we can say that the variables of electrons are denoted by

$$v = v_0 - \frac{e}{m}E(t_y)t_x, \tag{35}$$

$$z = v_0 t_x - \frac{e}{2m}E(t_y)t_x^2 \tag{36}$$

and

$$\rho = \frac{J_{peak}}{v} Max\left(\sin(\omega t_y), 0\right). \tag{37}$$

Let's assume that $E(ty)=E0\sin(\omega ty)$ which is synchronous to current modulation for the maximum interaction between electron beam and RF-field.

Then, we have

$$v = v_0 - \frac{e}{m}E_0 \sin(\omega t_y)t_x \tag{38}$$

and

$$z = v_0 t_x - \frac{e}{2m}E\sin(\omega t_y)t_x^2. \tag{39}$$

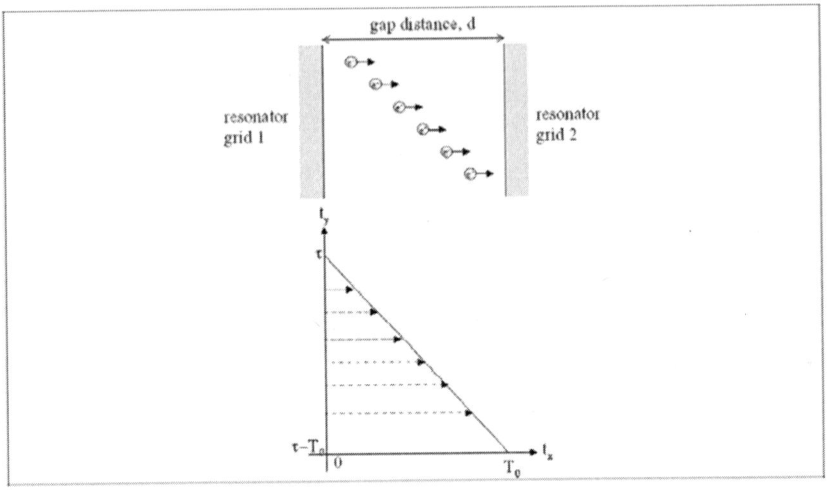

Fig. 10. Schematic representation for the definition of snapshot time ( ), transit time (*tx*) to *z*, departure time (*ty*)

By the way, from the above equation,

$$z = \frac{v_0 + v_z}{2} t_x \tag{40}$$

and

$$T = \frac{2d}{v_0 + v_d} \tag{41}$$

so that for (*period*)=167*ps*(λ=5*cm*)

$$T_0 \in \left[15ps, 30ps\right], \ \tau \in \left[0ps, 167ps\right], \tag{42}$$

and

$$t_x \in \left[0ps, 30ps\right]. \tag{43}$$

Figure 11 shows a typical case that electrons are decelerated due to the interaction between RF and electron beam. Extreme case would be 100

% donation of its kinetic energy to RF, which makes its velocity be zero at the resonator grid2. In that case electrons are delayed by 30 ps to the phase of RF field, where the half period of RF is 84 ps (6 GHz).

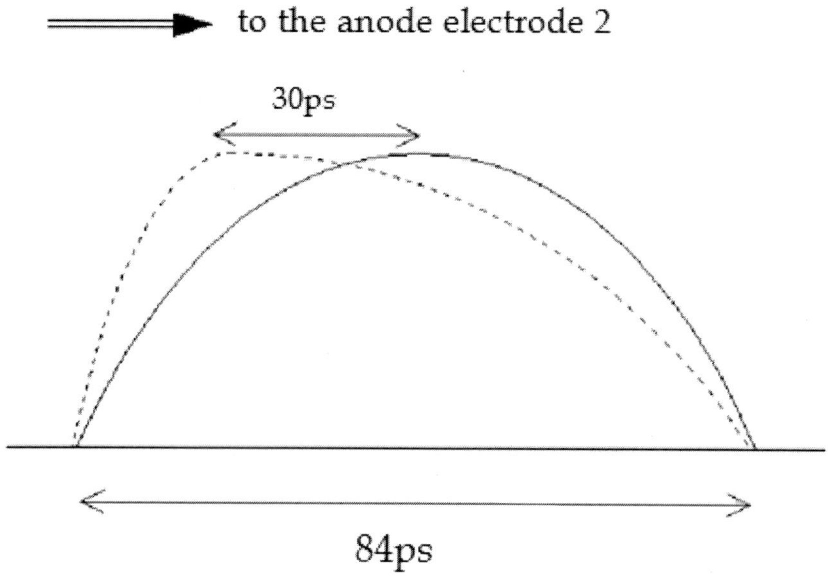

Figure 10. Electrons are decelerated due to the interaction between RF and electron beam. Extreme case would be 100% donation of its kinetic energy to RF, which makes its velocity be zero at the resonator grid2. In that case electrons are delayed by 30ps to the phase of RF field, where the half period of RF is 84ps (6GHz)

The resonator field theory is described by the equation

$$\frac{dU_E}{d\tau} = -\frac{\omega}{Q}U_E + P_{out}, \tag{44}$$

where *Pout* is the energy output from the bunch and *UE* is the electric energy in the cavity. The value of *Pout* depends on the behavior of the individual electrons as they move across the gap which in turn depends on the gap voltage and field profile. The axial electric field is only assumed by sinusoidal shape as *E*(*ty*)=*E*0sin(*ωty*)

Define

$$\langle f(t_y)\rangle \equiv \frac{\omega}{2\pi}\int_0^{\frac{2\pi}{\omega}} f(t_y)dt_y. \tag{45}$$

Then, the time-averaged output power becomes

$$\langle P_{out}\rangle = \left\langle \int_0^d \rho E v dz \right\rangle = \left\langle \int_0^d J_{peak} E_0 Max\left(\sin^2(\omega t_y), 0\right) dz \right\rangle$$
$$= \frac{\omega}{2\pi}\int_0^{\frac{2\pi}{\omega}}\int_0^d J_{peak} E_0 \sin^2(\omega t_y) dz dt_y = \frac{J_{peak} E_0 d}{4}. \tag{46}$$

Because E=0

when there are non electron charges, from the Maxwell's equation set,

$$\left\langle \frac{\omega}{Q} U_E \right\rangle = \frac{\varepsilon\omega}{2Q}\left\langle \int_0^d E_0^2 Max\left(\sin(\omega t_y), 0\right)^2 dz \right\rangle \cong \frac{\varepsilon\omega}{2Q}\frac{E_0^2}{2}\left\langle \int_0^d \sin^2(\omega t_y) dz \right\rangle$$
$$= \frac{\varepsilon\omega E_0^2}{4Q}\left\langle \int_0^d \sin^2(\omega t_y) dz \right\rangle = \frac{\varepsilon\omega E_0^2 d}{4Q}\left\langle \sin^2(\omega t_y) \right\rangle = \frac{\varepsilon\omega E_0^2 d}{8Q}. \tag{47}$$

Since

$$\langle P_{out}\rangle = \left\langle \frac{\omega}{Q} U_E \right\rangle \tag{48}$$

at steady state, we have

$$E_0 = \frac{2QJ_{peak}}{\varepsilon\omega}. \tag{49}$$

Therefore, the efficiency becomes that

$$\eta = \frac{\langle P_{out}\rangle}{\langle J\rangle V} = \frac{\frac{J_{peak} E_0 d}{4}}{\frac{J_{peak}}{\pi} V} = \frac{\pi E_0 d}{4V} = \frac{\pi Q J_{peak} d}{2\varepsilon\omega V}. \tag{50}$$

As an example,

$$\varepsilon = 8.85 \times 10^{-12} F / m,$$
$$\omega = 2\pi \times 10^{9} / \sec,$$
$$d = 0.4 \times 10^{-3} m,$$
$$Q = 37 \tag{51}$$

give us the followings

$$J_{peak} = \frac{3}{\pi\{(9.4\times10^{-3})^2 - (5.7\times10^{-3})^2\}} = 1.71[A / cm^2], \tag{52}$$

where V=2000*volts* e=1.6×10−19C, and m=9.1×10−31kg

Therefore,

$$\eta = \frac{\pi Q J_{peak} d}{2\varepsilon\omega V} = 0.60 \tag{53}$$

and

$$E_0 = \frac{2QJ_{peak}}{\varepsilon\omega} = 3.8\times10^6 V / m. \tag{54}$$

In the previous example, we have seen that the tube reveals 67% efficiency and the electric field in the gap region was 4×109V/m . This result of the numerical analysis is well matched with the above theoretical calculations.

On the other hand, we can compare MAGIC analysis with analytical calculation as can be seen Fig. 12 which shows efficiency vs. peak current density by MAGIC PIC simulation and analytical calculation. The slope of the line fitted with the results from MAGIC analysis shows 0.266.

From the analytical equation of efficiency, we can get the slope of the line, m, as the followings,

$$m = \frac{\eta}{J_{peak}} = \frac{\pi Q d}{2\varepsilon\omega V} = 0.348 cm^2 / A. \tag{55}$$

Analytical calculation says the slope of the line be 0.348 . MAGIC and analytical equation reveal that as we increase the current density, the efficiency of the tube also increases.

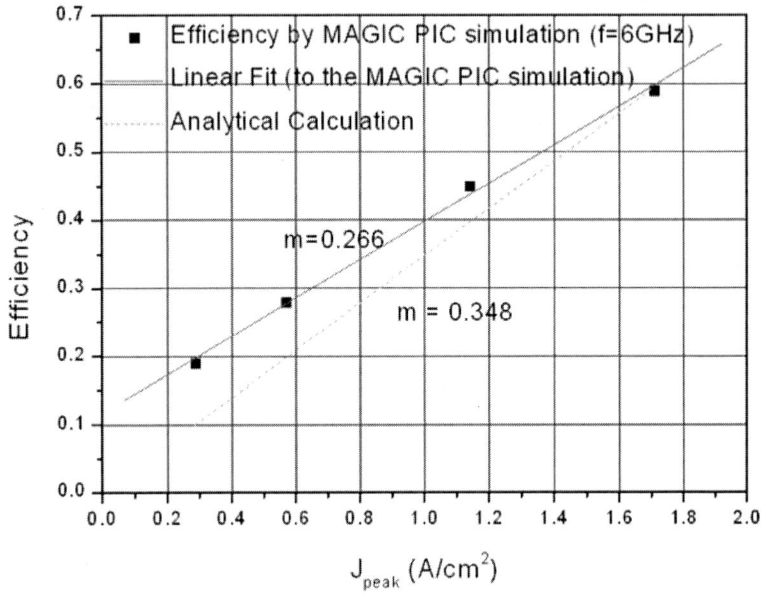

Figure 11. Efficiency vs. peak current density. Square dots are results from MAGIC simulation. Solid line is line fit with the results. Dotted line is analytical calculation. The slope of the solid line, line fit with the MAGIC results, is 0.266 and that of dotted line from analytical calculation is 0.348

4. CAVITY DESIGN FOR THE UNIFORM ATMOSPHERIC MICROWAVE PLASMA SOURCE

The atmospheric-pressure microwave-sustained plasma has aroused considerable interest for many application areas. The advantages of this plasma are electrode-less operation and efficient microwave-to-plasma coupling. It has the characteristics of high plasma density ($\approx 1013/cm-3$) and efficient energy conversion from microwaves to the discharged plasma which is high-efficient plasma discharge (~80%). An atmospheric-pressure microwave-sustained plasma can be formed in a rectangular resonant cavity, a waveguide, or a surface-effect system. This plasma has been widely used in the laboratory spectroscopic analysis, continuous emissions monitoring in the field, commercial processing, and other environmental applications.

Atmospheric-pressure microwave plasma sources consist of a magnetron as a microwave source, an isolator to isolate the magnetron from the harmful reflected microwave, a directional coupler to monitor the reflected microwave power, a 3-stub tuner to match impedance from the magnetron to that from the plasma generator, and a plasma generator through which gases pass and are discharged by the injected microwave. The plasma in the discharge generator is sustained in a fused quartz tube which penetrates perpendicularly through the wide walls of a tapered and shorted WR-284 waveguide.

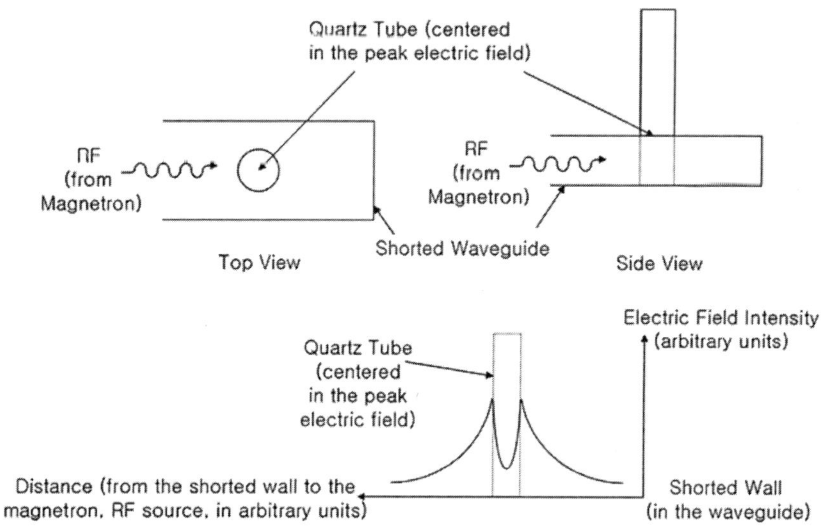

Figure 12. Schematic illustrations of the atmospheric pressure microwave plasma source using a quartz tube located a quarter wavelength from the end of the shorted waveguide to maintain a high electric field intensity to discharge the passing- gases to generate a plasma efficiently

The plasma's cross-sectional area of the atmospheric-pressure microwave plasma source is limited by the quartz tube's diameter, which is also limited by the maximum intensity of the electric field profile sustainable by the injected microwave and the shorted waveguide structure, as shown in the Fig. 13. The diameter of the quartz tube is set around 30 mm to maintain a dischargeable electric field intensity in the discharge area. If the diameter of the quartz tube becomes much larger than that, no discharge occurs. This is the main reason that the cross-sectional area cannot be extended more widely in the atmospheric-pressure microwave-sustained plasma source.

Here is an example of a rectangular reentrant resonator, a single ridge cavity, to overcome this size-limited plasma discharge in an atmospheric microwave-sustained plasma source. The box-type 915 MHz reentrant resonator has gridded walls on both sides of the gap so that the plasma gases pass thorough the grid holes and are discharged by the induced electric gap field due to the injected microwave energy. As we have studied in last section, it could be calculated analytically the cavity design formula by using lumped-circuit approximation, then, we performed a RF simulation by using the HFSS (High Frequency Structure Simulator) code. The designed plasma cross-sectional area is 810 mm x5 mm. The cavity has a fundamental TE_{10}-like mode to discharge gases uniformly between the gridded walls and has a 915 MHz resonant frequency. This large cross-sectional area plasma torch can be effectively used in commercial processing and other environmental applications.

4.1. Numerical analysis of the plasma source rectangular cavity

A reentrant cavity in cylindrical symmetry has been widely used in vacuum tubes such as klystrodes and klystrons. However, in order to keep the resonant frequency constant whatever the cavity lengths for atmospheric microwave plasma torch, it would be better to consider a rectangular reentrant cavity, as shown in Fig. 14. The resonant frequency of the cavity for the atmospheric microwave plasma source has a fundamental TE_{10}-like mode.

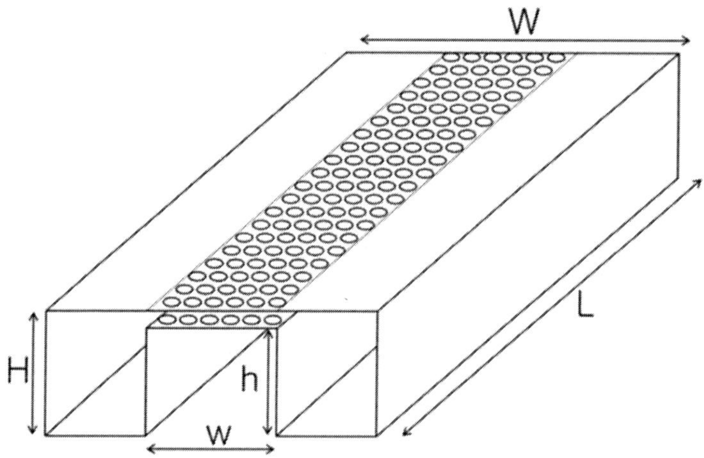

Figure 13. Schematic of the box-type resonator cavity as a atmospheric pressure microwave plasma source with grids to maintain a high electric field intensity to discharge the passing gases to generate plasma efficiently in the gap region

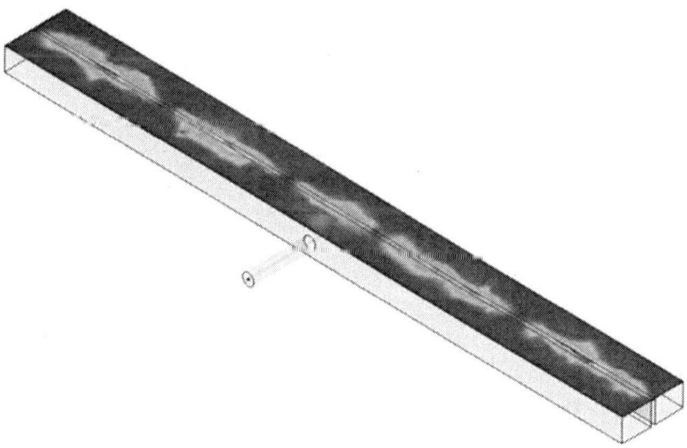

Figure 14. Electric field intensity profile of the TE10-like mode from the HFSS simulation on the mid-plane between the grid surfaces in the gap region. Both ends are shorted walls, and the electric field profiles are distributed within 810*mm*×5*mm*in the gap region. Microwave energy is feeding through the coaxial coupler to the cavity. If we use a pair of these cavities in parallel, the crest of one will compensate for the other's trough electric field intensity areas, which could result in a uniform microwave plasma source

The gap region between the two gridded walls in the reentrant cavity sustains a high electric field intensity and, thus, easily discharges the gases to the plasma state when it is excited by the microwave energy. With the grid planes surrounding the gap region, the TE_{10}-like mode is the dominant one, and 915 MHz is the cavity's resonant frequency. In the simulation process using the HFSS code, the cavity grids were assumed to be smooth conducting surfaces to design the box-type reentrant cavity.

In this example, the cavity for the microwave plasma source was set to be resonated at 915 MHz, rather than 2.45 GHz, so as to have a much larger cross-sectional plasma area. Its design, setting a cavity to 915 MHz, is based on the design formula obtained by using a theoretical calculation based on the lumped-circuit approximation for the cavity. Then, we used a simulation tool, the HFSS code, to check the cavity frequency.

From Gauss' law, the equivalent capacitance value of the reentrant cavity for a TE_{10}-like mode is

$$C = \frac{Q}{V} = \frac{Q}{E_{\perp}(H-h)} = \frac{\varepsilon_0 L w}{H-h}, \tag{56}$$

where Q, V, and $E\perp$are the stored electric charges on the grid surfaces, the electric voltage difference, and the perpendicular electric field intensity between the opposite grid surfaces respectively, and w, H, and $h$are designated dimensional parameters, as shown in Fig. 14. From Ampere's law, the equivalent inductance value of the reentrant cavity for a TE_{10}-like mode can be expressed as follows. Since the magnetic field flux density in the cavity is $B=\mu 0I/2L$, and since the total magnetic flux in the box-type reentrant cavity is $\Phi=BH(W-w)$, from the inductance definition,

$$L = \frac{\mu_0 H(W-w)}{4L} . \tag{57}$$

Therefore, the cavity resonant frequency is given by

$$f = \frac{c}{\pi}\sqrt{\frac{H-h}{Hw(W-w)}} \; . \tag{58}$$

This resonant frequency formula shows that the resonant frequency is independent of the cavity length *L*, which means that if we extrude the plasma through this grid planes, the plasma's cross-sectional area can be uniformly extended as much as we want without any influence on the cavity resonant frequency. However, because we use an available microwave source, which has a limited RF power, to discharge the gases, the cavity length cannot be extended to an unlimited extent. We will discuss this relationship between the consumption RF power and the cavity length in the next paragraph.

Since we have a theoretically-approximated expression for the design parameters, we should use this formula to pick up the initial design parameters to get detailed parameters for the resonator reentrant cavity as a plasma source. Then we should investigate the RF characteristics using the design tool, HFSS.

The energy stored in the gap region can be expressed as

$$U = \frac{1}{2}CV^2 \; , \tag{59}$$

where *C*and *V*are the equivalent capacitance value of the cavity and the induced gap peak voltage between the grid planes, respectively. If we assume that all of this stored energy can be delivered to the gas discharge reaction process to generate the plasma, the microwave power consumption can be expressed as

$$P = \frac{1}{2} fCV^2 = \frac{LV^2}{2\pi}\sqrt{\frac{\varepsilon_0 \omega}{\mu_0 H(H-h)(W-w)}} \; . \tag{60}$$

In other words, the cavity length for the atmospheric-pressure microwave-sustained plasma source is

$$L = \frac{2000\pi P}{V^2}\sqrt{\frac{\mu_0 H(H-h)(W-w)}{w}} \; , \tag{61}$$

where *V*is in kV, *P*is in kW, and the dimensional parameters, *L*, *H*, *h*, *W*, and *w*, are in *mm* units in this formula.

Fig. 15 and Fig. 16. show a rectangular cavity design example using one of the available magnetrons, 915 MHz, 60 kW, as a RF source, and the geometrical parameters to *H*=29*mm*, *H*−*h*=1*mm*, and *W*=80*mm*. Furthermore, if the gap voltage is the upper-limit RF air breakdown voltage, that is, the DC breakdown voltage for air at the level of 33*kV*/*cm*at atmospheric pressure, the cavity length is limited by 810*mm* for self discharge. In this case, the plasma's cross-sectional area is uniformly distributed within 810*mm*×5*mm* , i.e., *w*=5*mm*.

This uniformly distributed plasma area is 810*mm*×5*mm*, which is much larger than the areas of the conventional atmospheric-pressure microwave-sustained plasma sources. This large-sized microwave plasma at atmospheric pressure can be used in commercial processing and other environmental applications area.

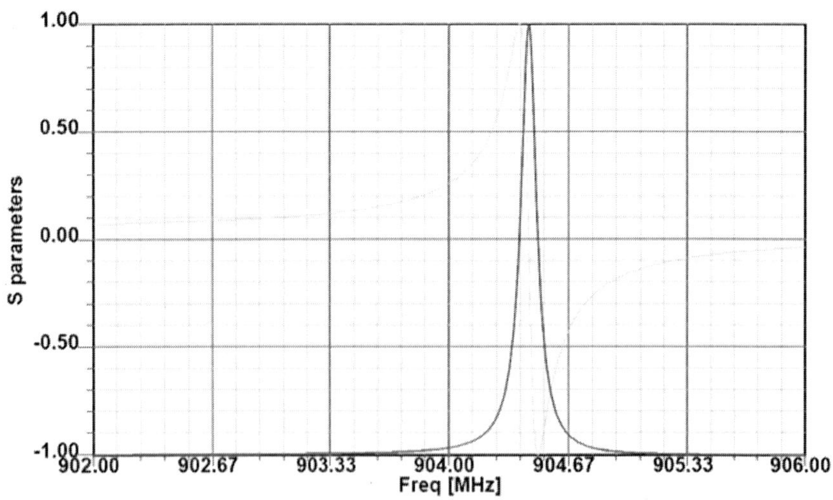

Figure 15. Scattering parameters of the box-type reentrant cavity. The solid red and dashed blue lines are the real and imaginary values of S11, respectively. This shows that the cavity has a resonant frequency of 904 MHz and a high Q value of 7400

5. CONCLUSION

The annular beam cavity design was investigated analytically and simulated using the HFSS and MAGIC PIC codes to find the fine-tuned design parameters and optimum efficiency of the TM_{01}-mode operation in the klystrode with the reentrant interaction cavity. We also studied how to induce the governing efficiency formular for the microwave vacuum tube, klystrode. The efficiency of this exampled annular beam klystrode reveals 67 % at 6.004 GHz. The theoretical calculation anticipated that the efficiency would be 60% and the electric field intensity 4000V/m which is well matched with the results of numerical analysis. This concept could yield an economic sized device comparable to commercial magnetron devices.

A single ridge cavity was designed to show how to design uniform atmospheric microwave plasma source which does not need ignitors for the initial discharge. The single ridge cavity in the shape of resonator was built in the grid structures for the wide cavity torch of atmospheric pressure microwave plasma. This cavity design process has been studied by analytical calculation using lumped circuit approximation and simulation using commercial 3D HFSS code. A self-dischargeable study also has been investigated for the breakdown feasibility in the cavity gap region. These cavities show the uniform electromagnetic wave distribution between the grids through which large-sized microwave plasma could be generated.

REFERENCES

1. Ansoft Co. 1999 High Frequency Structure Simulator, version 6.0, Ansoft Co., PA, USA, 1999
2. A. J. Burton, G. F. Miller, 1971 The application of integral equation methods to the numerical solution of some exterior boundary-value problems, Proc. Roy. Soc. Lond., A. 323, 201 EOF

3. H. J. Curnow, 1965 A General Equivalent Circuit for Coupled-Cavity Slowe-Wave Structures, IEEE Transactions on Microwave Theory and Techniques, MTT-13 , 5 671

4. F. Dai, A. S. Omar, 1993 Field-analysis model for predicting dispersion property of coupled-cavity circuits, IEEE MTT-S Digest, 901

5. K. Fallgatter, V. Svoboda, J. D. Winefordner, 1971 Physical and analytical aspects of a microwave excited plasma, Applied Spectrocopy, 25, 3 347 EOF 352 EOF

6. K. Fujisawa, Fujisawa 1951 Theory of slotted cylindrical cavities with transverse electric field, Tech. Reps. Osaka Univ., 1 69

7. K. Fujisawa, 1958 General treatment of klystron resonant cavities, IRE Trans., MTT-6, 344 EOF 358 EOF

8. A. S. Gilmour Jr, 1986 Microwave Tube, Artech House, Chap. 15

9. B. Goplen, L. Ludeking, K. Nguyen, G. Warren, 1990 Design of an 850MHz klystrode, International Electron Devices Meeting, 889

10. B. Goplen, L. Ludeking, D. Smithe, G. Warren, 1994 MAGIC User's Manual, Rep. MRC/WDC-R-326, Mission Res. Corp., Newington, VA

11. D. R. Hamilton, J. K. Knipp, J. B. Horner Kuper, 1966 Klystrons and Microwave Triodes, Dover Pub., New York

12. W. W. Hansen, 1938 A type of electrical resonator, Journal of Applied Physics, 9 654

13. Be. Z. He, 1997 The cold quality Qcold for magnicon on resonator in the rotating TMn10 mode, International Journal of Infrared and Millimeter Waves, 18 2 437

14. Hyoung. S. Kim, Han. S. Uhm, 2001 Analytical Calculations and Comparison with Numerical Data for Annular Klystrode, IEEE Transactions on Plasma Science, 29 6 875 880

15. Kim, Hyoung Suk & Ahn, Saeyoung 2000 Numerical Analysis of C-band Klystrode with Annular Electron Beam, International Journal of Infrared and Millimeter Waves, 21 1 11 20

16. R. E. Kleinman, G. F. Roach, 1974 Boundary Integral equations for the three-dimensional Helmholtz equation, SIAM Review, 16 2 214 EOF

17. S. W. Lee, H. J. Kang, H. S. Kim, H. S. Hum, 2009 Numerical Design of the Cavity for the Uniform Atmospheric Microwave Plasma Source, Journal of Korean Physical Society, 54 6 2297 2301

18. Samuel. Y. Liao, 1990 Microwave Devices & Circuits, Prentice Hall

19. R. B. Marks, 1986 Application of the singular function expansion to an integral equation for scattering, IEEE Transactions on Antennas and Propagation, AP-34 , 5 725 EOF 728 EOF

20. S. W. Mcdonald, J. M. Finn, M. E. Read, W. M. Manheimer, 1986 Boundary integral method for computing eigenfunctions in slotted gyrotron cavities of arbitrary cross-sections, Int. J. Electronics, 61 6 795 EOF 822 EOF

21. L. W. Pearson Pearson, 1984 A Note on the representation of scattered fields as a singularity expansion, IEEE Transactions on Antennas and Propagation, AP-32 , 5 520 EOF 524 EOF

22. A. G. Ramm, 1982 Mathematical foundations of the singularity and eigenmode expansion methods, Journal of Mathematical Analysis and Applications, 86, 562 EOF 591 EOF

23. R. J. Jr Riddell, 1979 Boundary-Distribution Solution of The Helmholtz Equation for a Region with Corners, Journal of Computational Physics, 31, 21 EOF 41 EOF 1979

24. C. Riekmann, A. Jostingmeier, A. S. Omar, 1996 Application of the eigenmode transformation technique for the analysis of planar transmission lines, IEEE MTT-S Digest, 1023

25. Samuel Seely, 1950 Electron-Tube Circuits, McGraw Hill Book Com. Inc., New York

26. A. F. Stevenson, 1948 Theory of slots in rectangular wave-guides, Journal of Applied Physics, 19 24,

27. J. Tan, G. Pan, 1996 A general functional analysis to dispersive structures, IEEE MTT-S Digest, 1027 EOF

28. W. Tobocman, 1984 Calculation of acoustic wave scattering by means of the Helmholtz integral equation. I, J. Acoust. Soc. Am., 76 (2), 599 EOF

29. A. T. Zander, G. M. Hieftje, 1981 Microwave-supported discharges, Applied Spectroscopy, 35, 4 357 EOF 371 EOF

Chapter 2

On The Use of Electromagnetic Waves as Means of Power Supply in Wireless Sensor Networks

Jorge Cortés-Sánchez[1], Andrés Velázquez-Ramírez[1], Andrés Lucas-Bravo[1], Mario E Rivero-Angeles[2]*and Victor A Salinas-Reyes[1]

[1]*Telematics Department, UPIITA-IPN, Mexico City 07340, Mexico.*

[2]*Communication Networks Laboratory, CIC-IPN, Mexico City 07738, Mexico*

ABSTRACT

Wireless sensor networks (WSNs) can be typically used to achieve continuous monitoring (CM) or event detection inside the supervised area. In CM applications, each sensor node transmits periodically its sensed data to the sink node, while in event-detection driven (EDD) applications, once an event occurs, it is reported to the sink node. Hence, CM applications entail much higher energy consumption since all nodes are actively transmitting information much longer than typical EDD applications. Furthermore, in highly dense WSNs, the energy consumption is even higher. As such, the use of autonomous devices that provide a constant energy supply is becoming relevant for these environments. In this work, we propose to take advantage of the electromagnetic waves found in the radio-electric spectrum in order to supply the energy to each node in the network. To this end, the antenna and the storage and amplifications system for such device are designed. Additionally, the performance of dense CM WSN is studied using a Markov chain in order to calculate the lifetime of the system.

Keywords: Antenna design Markov analysis Voltage multiplier Wireless sensor networks

1 INTRODUCTION

Wireless sensor networks (WSNs) are typically designed for either continuous monitoring (CM) or event-detection driven (EDD) applications. EDD WSNs are deployed over a target area to supervise certain phenomena of interest. Once an event occurs, it is reported to the sink node by the sensors within the event area. Each node takes readings from the local environment and processes and transmits the sensed data to the sink node. In this type of WSNs, communications are only triggered by the occurrence of a pre-specified type of events. As opposed to EDD applications, CM WSNs are deployed in order to examine the evolution of certain parameters, which are refreshed periodically at the sink node. As such, CM applications have typically much higher energy. Energy consumption is a major issue in WSNs (for both EDD and CM applications) since in many cases it is very complicated to replace the node's batteries or even impossible. Consider for example, the case of a WSN deployed in the forest for animal protection purposes [1] or inside the concrete in buildings or steel beams in bridges in order to monitor the structural health [2].

In the literature, it is common to assume that nodes are equipped with batteries, which power the nodes for a limited amount of time [3]. Hence, the communication protocol is designed in order to limit the power consumption in the system. For instance, clustering protocols like the one proposed in [4] aim at reducing and evenly distributing the energy consumption in order to extend the network's lifetime.

Building on this, it is natural to search for alternatives to constantly supply power to the nodes in order to extend the network lifetime or even avoid complete energy depletion. Emerging technologies such as solar cells can be used for such effect with the main drawback that it cannot work efficiently indoors or at night. Another alternative would be the use of devices that take the energy from the electromagnetic waves in the radio-electric frequency spectrum, such as TV and radio signals. This alternative has the big advantage that, in these frequencies, there are almost always data transmissions for where to extract energy. In this case, the proposed device is not designed to decode these signals, just to take advantage of the ever-present signal transmission in this frequency range.

In this work we focus on the later alternative. To this end, we design both the antenna to receive electromagnetic waves in a wide frequency range that comprises radio and TV signals as well as the storage and amplification systems that provide a constant power supply to the nodes in the network. Figure 1 shows the proposed energy-harvesting system. An incident power of the electromagnetic waves is absorbed by the effective area of the antenna and passes through a transmission line to reach a voltage multiplier. Then the peak output voltage is increased four times the peak input voltage. The output of this stage corresponds to an alternating current (AC) quantity. Hence, the voltage multiplier output has to be processed by a power converter to produce a suitable direct current (DC) output voltage to meet the requirements of the application. Hence, the received power is rectified to eliminate AC components.

As an additional feature of our work, we develop a mathematical analysis to determine the performance of a CM WSN when such device is used, i.e., when nodes use a constant power supply. We focus our attention in CM WSNs since these applications are more energy consuming than their EDD WSNs counterpart.

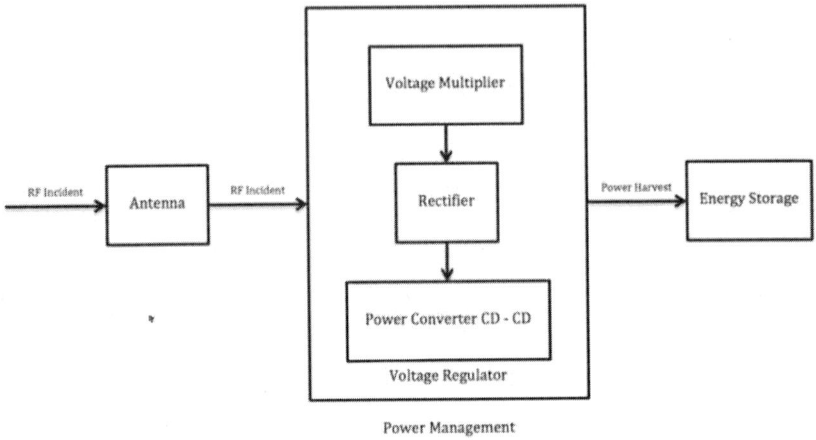

Figure 1 System block diagram.

From the numerical results obtained in this work, it is clear that the use of a constant power supply, such as the device proposed in this paper, can radically change the analysis and design of WSNs. Indeed, the design of WSNs is mainly focused on the lifetime of the system when a limited-energy battery is used as means of power supply. Specifically, communication protocols (including random access strategies and routing schemes) are mainly designed in order to limit the amount of transmissions. This leads to a very restrictive environment where only the most relevant data is to be transmitted. Conversely, when each node has a constant power supply, the design can now be focused on other parameters such as delay guarantees in multimedia data transmission, such as video surveillance or video transmission and VoIP-enabled sensors for noise detection.

A similar RF energy-harvesting device was presented in [5]. However, in [5] only a 2.4-GHz frequency was considered. In this work a range of frequencies (from some megahertz to a few gigahertz) are contemplated in order to harvest the maximum amount of energy from different radio sources. Also, the work in [5] does not present a performance evaluation of a WSN with a constant energy source since its focus is mainly on the electronic device for general purpose. In this line, in [6, 7] an alternating current/direct current (AC/DC) converter from an AC voltage of a few hundred millivolts is developed. In the work presented in [6], authors consider the use of a low-voltage AC/DC converter system for the particular case of a WSN. From the developed systems in [6, 7], it is clear that their proposed devices can indeed offer a constant power supply to the nodes in the system. However, both [6, 7] focuses mainly on the AC/DC conversion. Hence, neither antenna design is proposed nor the performance analysis of the WSN is considered.

In the rest of the paper, we develop the antenna design in order to harvest energy in a wide range of frequencies. Then the electronic device that converts the very low AC signal to a 3.5-V DC signal is presented along with the storage system. Finally, the performance analysis of a continuous monitoring WSN when nodes have a constant power supply is shown. We conclude the paper with some remarks and future work ideas.

2 REVIEW

In this section, the antenna design and the voltage multiplier are presented. As a case of study, it is considered that the WSN is deployed at a distance of 15 km from the radio and TV broadcast emissions. Also, a performance analysis of the proposed device in a WSN for continuous monitoring applications is presented.

2.1 Antenna selection

In order to take advantage of the ever-present signals in the radio and TV frequency spectrum, an antenna with a wide bandwidth is required. As such, a spiral antenna was selected. In Figure 2 we present the architecture of such antenna. Since the system was designed to operate in Mexico City, the following frequencies are selected: radio services in the range of 88 to 108 MHz with a received power of -37 dBm in average; TV services in the range of 170 to 700 MHz with a receive power of -24 dBm in average, as well as the range of 2.4 to 2.4835 that corresponds to the WLAN (WiFi) services which are experimenting an important growth in Mexico City. It can be seen in Figure 3 that in these selected frequencies, the antenna has an acceptable gain with small variations.

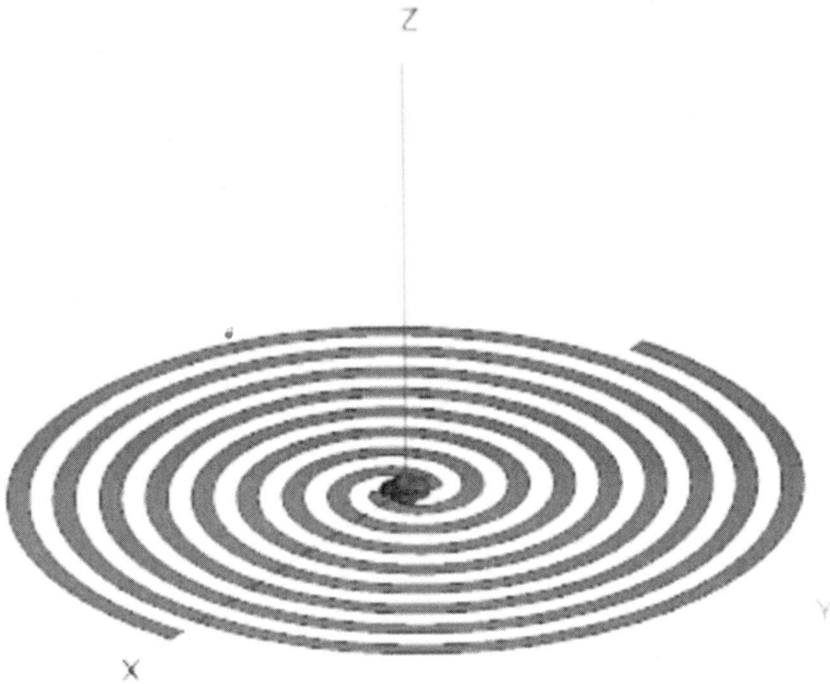

Figure 2 Spiral antenna structure.

The geometrical data for the antenna is as follows:

- 'Antenna diameter, 0.56 m
- 'Wire radius, 0.00875 m
- 'Separation between spires, 0.0175 m

In Figures 4 and 5, we show the radiation pattern for different angles of the antenna, which is stable and high for different frequencies, specifically in the frequency range of interest.

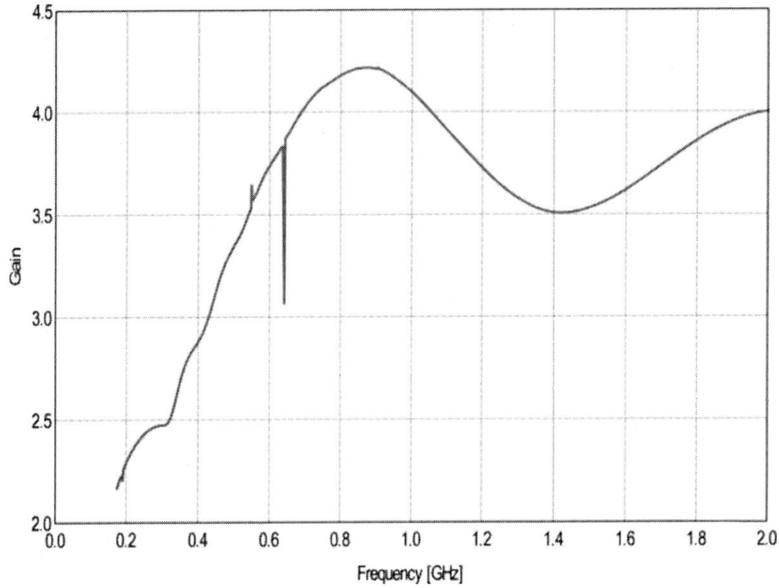

Figure 3 Antenna gain (times).

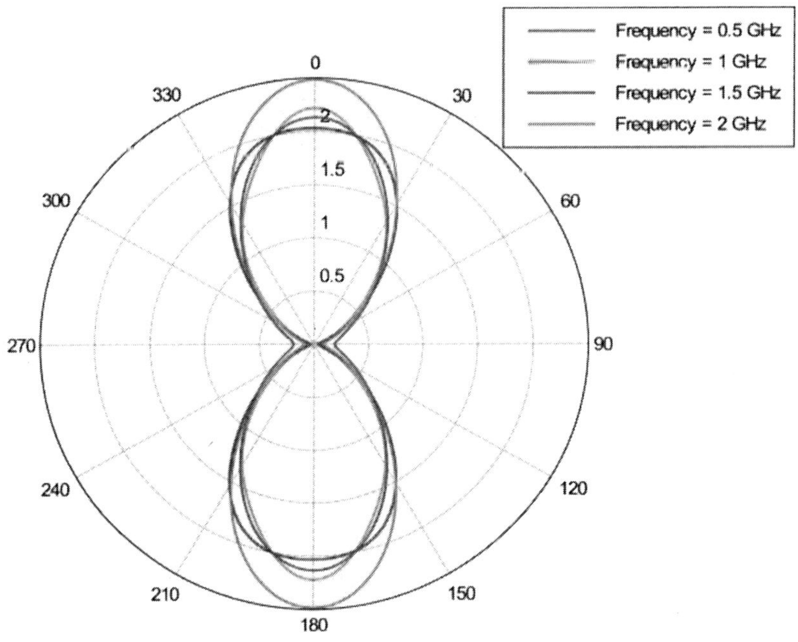

Figure 4 Radiation pattern at 0° (dBs).

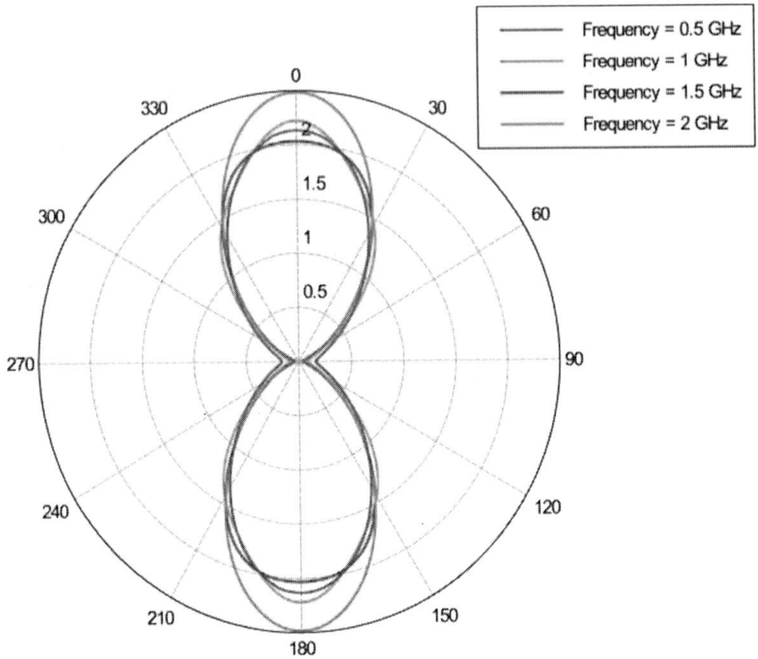

Figure 5 Radiation pattern at 90° (dBs).

One main reason for the antenna selection is its input impedance response which is also very stable in the operation range (300 MHz to 3 GHz). This implies a good coupling with the receptor. This is the main reason to use this type of antenna. Indeed, many other antennas were investigated, but the input impedance of a spiral antenna is predominantly real over a wide bandwidth (Figure 6) [8].

The spiral antenna can be used over a wide range of RF incident power densities, where the goal for the power management is to optimally load the rectenna at the DC output to achieve maximum harvested power [5]. The spiral antenna can be connected to switch with reconfigurable rectenna circuit. The characteristic of the antenna was simulated with FEKO software.

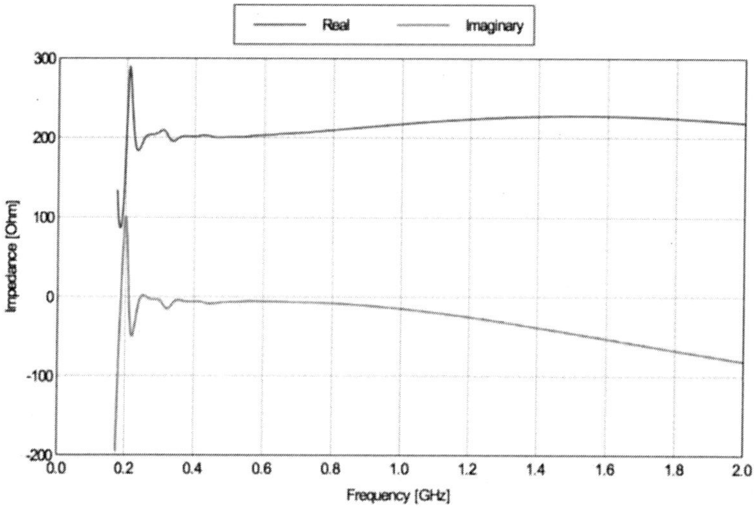

Figure 6 Antenna impedance (ohms) for the selected frequency range.

The RF input energy is rectified using a rectenna circuit. The circuit is optimized for a given input RF power. The principal limitation of the rectenna circuit is that it is designed for an operating point. Good efficiency is obtained when different parameters are considered: power level, central frequency, and load impedance [9].

A rectenna is a microwave rectifier which converts RF energy directly into DC current. When only one frequency is considered, it is possible to use a high-gain receiving antenna.

2.2 Storage and amplification

In this section, the electronic design for the voltage multiplier is presented. This system is shown in Figure 7.

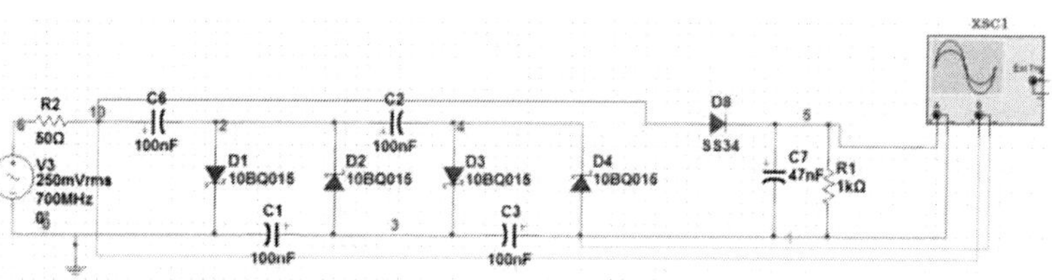

Figure 7 General storage and amplification system.

The antenna is considered to be an AC voltage source as seen in Figure 8a. It is then connected to the storage and amplification system via a transmission line with an impendence of 50Ω. Then, the signal enters to an amplification phase that increases the amplitude of the signal four times the value received at the antenna. Finally, at the end of Figure 7, a rectification system is used in order to have a DC signal with a constant value of 3.5 V, as seen in Figure 8b.

a

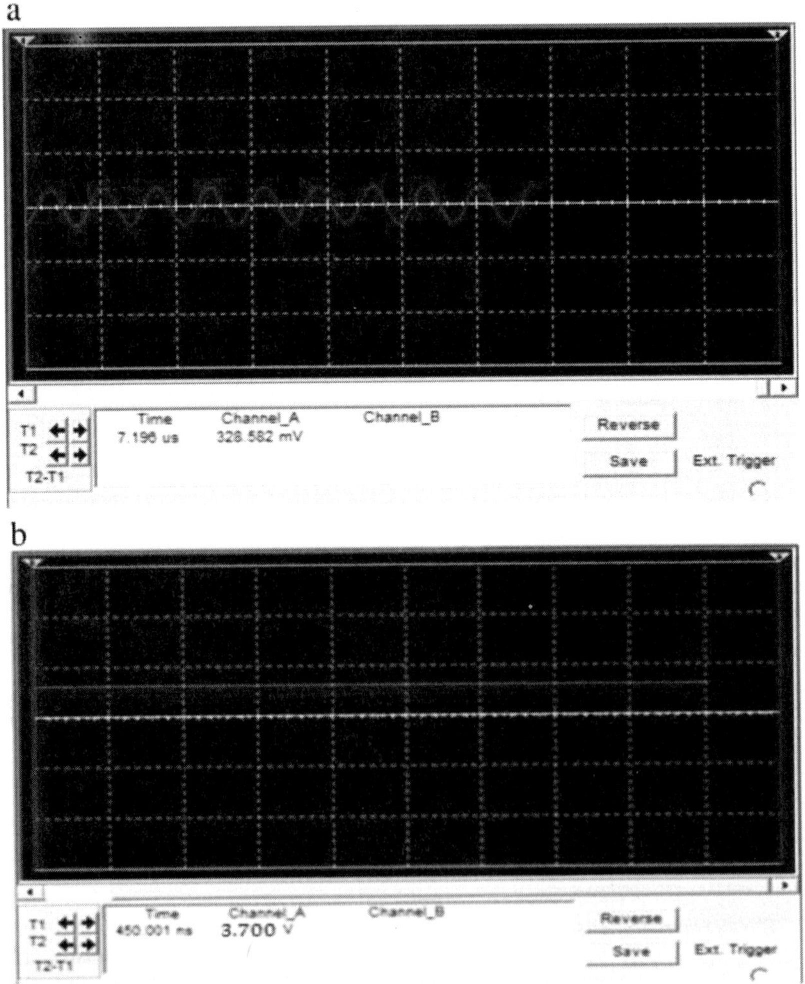

Figure 8 Input (a) and output (b) signals from the storage and amplification stage.

2.3 Analysis of the CM WSN

In this section, the advantages of using a device that delivers a constant power supply to each sensor node are studied.

It has been proven in [10] that the use of a clustered-based architecture greatly reduces the energy consumption in CM WSNs since the main sources of energy wastage are addressed: collisions, overhearing (when a node receives an unintended packet), idle listening (lost energy while listening to the medium to receive possible traffic that is not sent), and overhead (due to exchange of signaling messages required for the protocol execution). Hence, we assume such architecture in this analysis.

In a clustered-based network, like LEACH [4], two distinct phases compose the normal operation: cluster formation (CF) and steady state as shown in Figure 9. In the CF phase, nodes transmit to the rest of the nodes using a random access protocol such as slotted CSMA-NP, while in the steady-state phase (when clusters are already formed, a cluster head is selected for each cluster and the cluster members are well identified), a collision-free protocol can be used such as a TDMA-based structure. Since the role of the cluster head (CH) is more energy consuming than the operation of the cluster members (CMs) (due to the data recollection and transmissions to the sink), the function of the CH has to be periodically interchanged in order to avoid premature energy depletion of a node. The time period that comprises the CF and the steady state (with multiple TDMA frames) is known as a round. As such, the energy consumption in a CM WSN has to consider both phases. Multiple reports are transmitted inside a round, where CMs send their information to the CHs in their specific time slot. When all the CMs have transmitted to the CH, the CH sends this information to the sink node also in a specific time slot. Specifically, we are interested on studying the system's performance in terms of the number of nodes that causes the energy depletion of a given node before the end of a round. Indeed, when the number of nodes is small and when a constant power supply is used, none of the nodes in the network consumes all of its energy. This is because both the CF and steady-state phases do not require the use of more energy than the one delivered by the proposed device. In such case, the analysis is straightforward since nodes never consume all their energy. Conversely, when the number of nodes is high, the number of collisions and the time spent in idle listening and overhearing is considerable. As such, it is likely that some nodes may deplete their energy before a round is finished. This is a very undesirable behavior in a WSN since the premature energy depletion of a node implies a loss of information. Building on this, we focus our analysis to find the number of nodes that produces information loss in a WSN when the constant power supply is both used and not used.

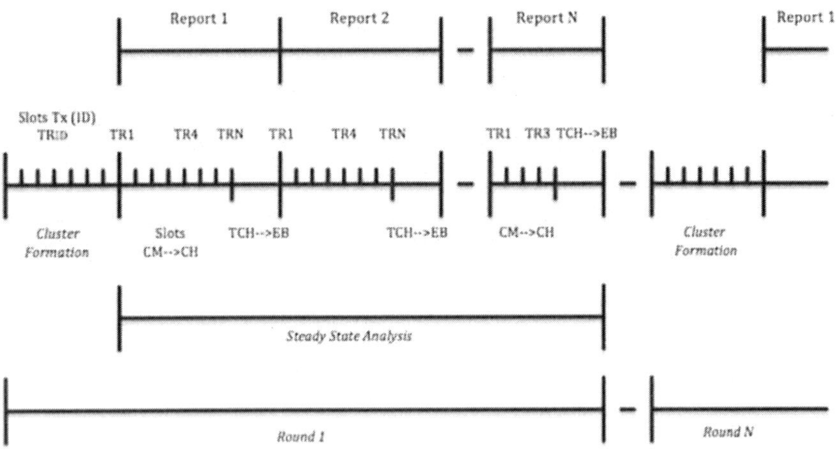

Figure 9 Data transmission in cluster-based WSNs.

The following parameters are used for the analysis: The WSN is considered to be placed at 15 km from the emitting antenna. In such conditions, the constant power supply is 3.5 V. The control packets used for the CF phase is $l_{CF} = 80$ b, while the data packets used in the steady phase is $l_d = 2{,}048$ b. Based on commercial transceivers (Microchip MRF24J40, Microchip Technology Inc, Chandler, AZ, USA), the transmission energy per bit is $E_{Tx} - 0.395$ J while the reception energy per bit is $E_{Rx} = 82.5$ mJ. Also the energy of a conventional lithium battery is 7.59 J.

For a WSN with N nodes and C clusters in average, there are $E_{NC} = N/C$ nodes per cluster.

2.3.1 Steady-state analysis

Let us first consider the energy consumed at the steady state. In this case, the E_{NC} nodes in the cluster transmit a data packet to their respective CH. Also, the CH transmits all the gathered data in the cluster to the sink of the network. As such, the energy consumed at the steady state is

$$E_s = \left[E_{Tx}\, l_d\left(\frac{N}{C}-1\right) + E_{Tx}\, l_d\left(\frac{N}{C}\right)\right](R), \tag{1}$$

where l_d is the data packet length used for the transmission of the recollected data by each node, and R is the number of TDMA frames per round. Note that from the N/C nodes in average that form the cluster, only the cluster members transmit to the CH (this comprises the first two elements of (1)). Also, the CH transmits all the recollected packets to the sink node, including its own data packet. As such, neither data compression nor aggregation is assumed in this analysis. Also, it is considered that the same transmission energy is used to reach the CH and the sink node. The rationale behind this is that the analysis presented in this work is a first attempt at analyzing the performance of a WSN with a constant power supply. To this end, we consider the worst conditions. The effect of data compression and/or aggregation would render less energy consumption.

2.3.2 Cluster formation analysis

Now we focus on the cluster formation phase. This process can be modeled by a two-dimensional transitory Markov chain with states (n,m) where n $(0 \le n \le N)$ corresponds to the number of nodes attempting to transmit their control packet while m $(0 \le m \le n)$ is the number of transmissions in a given slot as depicted in Figure 10.

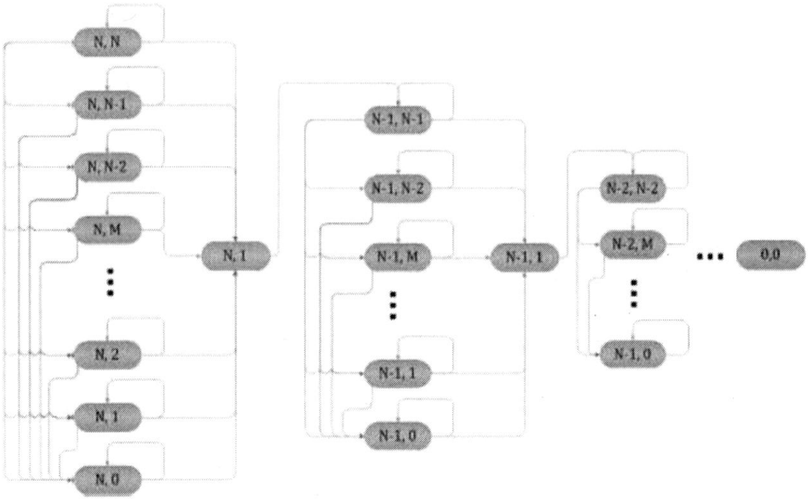

Figure 10 Markov chain for the cluster formation phase.

The system begins in state (*N*,0) and state (0,0) is an absorbing state (in this state all the nodes in the system have successfully transmitted their control packet and the clusters are formed).

From any state (*n*,*m*), the system can go to any other state (*n*,*l*) $0 \le l \le n$ with probability:

$$P_{mn,ml} = \binom{n}{l} \tau^{l} (1-\tau)^{n-1}, \tag{2}$$

where τ is the transmission probability given by the geometric backoff assumed in this work. When the chain enters state (*n*,1) (one node form the *n* contending nodes), it goes to state (*n*-1,l) with probability 1/(*n*-1).As such, this two-dimensional Markov chain can be considered to be composed of *N* different phases denoted by the number of nodes attempting a packet transmission. Hence, the energy cxonsumption at phase *n* can be calculated noting that whenever a successful transmission occurs, there is one node that consumes $E_{\mathrm{Rx}}^{\mathrm{CF}} = E_{\mathrm{Tx}} l_{\mathrm{CF}}$ units of energy, while there are *n*-1 nodes that receive the packet, each consuming $E_{\mathrm{Rx}}^{\mathrm{CF}} = E_{\mathrm{Rx}} l_{\mathrm{CF}}$ units of energy. Hence, the energy consumption in the case of a successful transmission is . $E_{\mathrm{Tx}}^{\mathrm{CF}} + (n-1)E_{\mathrm{Rx}}$. On the other hand, whenever a collision occurs or if there are no transmissions, there are S_n nodes that transmit, each one consuming $E_{\mathrm{Tx}}^{\mathrm{CF}}$ units of energy, while n-S_n nodes listen to the channel consuming $E_{\mathrm{Rx}}^{\mathrm{CF}}$ units of energy each one.

As such, the cost associated to go from state (*n*,*m*) to state (*k*,*l*) can be denoted as

$$C_{nm,kl} = lE_{Rx}^{CF} + (n-l)E_{Rx}^{CF}. \tag{3}$$

Let us denote by C_{nm} the cost of being in state (*n*,*m*). It can be expressed as

$$C_{nm} = \sum_{l=0}^{n} C_{nm,kl} P_{nm,kl}. \tag{4}$$

As such, the energy consumption form state (*n*,*m*) to the absorbing state (0,0) can be calculated as [11]

$$EC_{nm} = C_{nm} + \sum_{k=0}^{n} \sum_{l=0}^{k} P_{nm,kl} EC_{kl} \tag{5}$$

Hence, the average energy consumption at the cluster formation phase is $EC_{N\,0}$.In Figure 11, the total energy consumption per node and per round is presented with $R = 1$. In Figure 11, two rectangles are shown. In the red one, we show the operation range of a WSN where nodes are powered by a conventional battery. In the black rectangle, the operation range of a WSN with a constant power supply, such as the one proposed in this work, is shown. It can be seen that as the number of nodes in the system increases, also the energy consumption in the CF phase increases due to the number of collisions, idle listening, and overhearing. Also, in the steady state, the energy consumption increases since there are more nodes reporting its information to the CH. As such, CHs consume much more energy in order to transmit the gathered data to the sink node. It can be seen that without the proposed device, the system only support 40,000 nodes without a node depleting its energy before the first round is over. On the other hand, when the device is used, the system supports 140,000 nodes before a node depletes its energy before the first round is over, which corresponds to an increase of 10 times the number of nodes supported in the highly dense WSN.

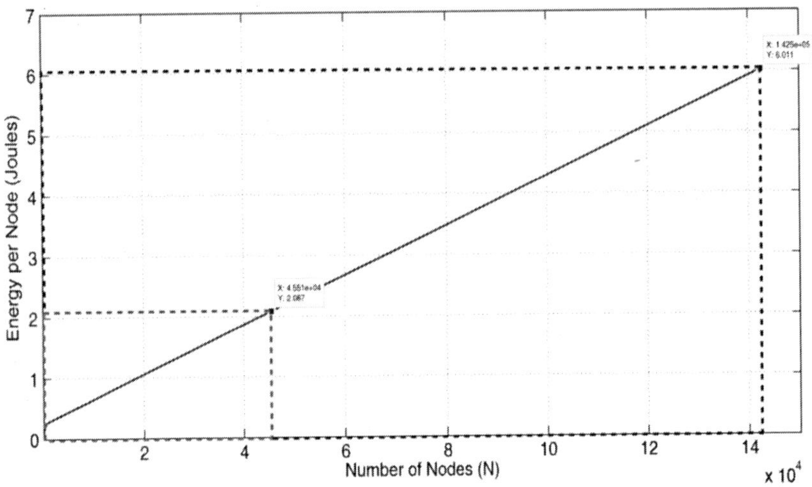

Figure 11 Total average energy consumption per node in a round for different number of nodes in the system.

Figure 12 shows the lifetime of the system for different number of nodes in the network. The cases where nodes are powered only with batteries and when a constant power source is used are presented. Whenever the number of nodes is small, the lifetime of the system is high since there is a low energy consumption. However, as the number of nodes increases, the system's lifetime decreases rapidly. However, for any number of nodes, the system's lifetime is always much higher when a constant power source is used. This result justifies the use of such a device in any environment.

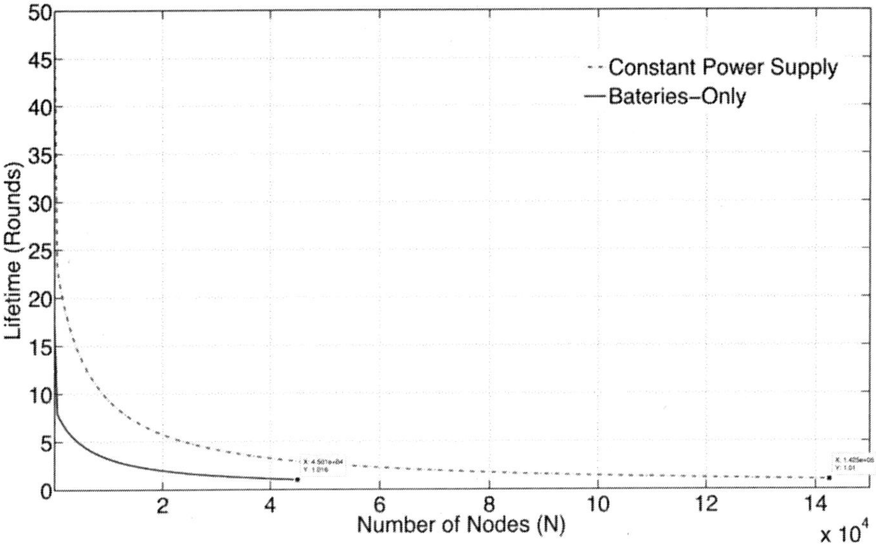

Figure 12 System lifetime for different number of nodes in the network.

3 CONCLUSIONS

In this paper an electronic device to use the energy from the electromagnetic waves was designed. First, the main radio emissions in the particular case of Mexico City were studied in terms of frequency and RF power reception. Then, a suitable antenna was designed such that it is capable of absorbing most of the available radio emissions ranging from dozens of megahertz to a few gigahertz. Given this wide range of frequencies of interest, it can be connected to switch with reconfigurable rectenna circuit. As such, a spiral antenna was selected given its good coupling characteristics in practical all the frequency range. Then, the electronic system was designed. This circuit is composed of the AC/DC converter, rectifier, amplification, and a storage system that directly powers each sensor node.

Building on this, a mathematical analysis to calculate the energy consumption in a CM WSN was performed. First, the analysis considered that a limited energy battery powered the nodes in the system. This is the conventional case in the WSN environment. Then, the analysis was performed considering a constant power supply of

3.5 V. Hence, whenever the energy consumption for the continuous monitoring application entails an energy consumption of less than 3.5 V per node, the system can operate without suffering any depletion from the nodes. Conversely, when the energy consumption is higher than this value, some nodes stop operating, causing information losses. This simple analysis proves the benefits of the proposed device in the context of WSNs increasing by 10 the number of nodes supported in the system without a node depleting its energy in the first round. The analysis can be further improved by considering multiple TDMA frames per round as well as compression and aggregation techniques. Also, different transmission ranges can be considered in order to further reduce the energy consumption.

DECLARATIONS

Acknowledgements

This work was partially supported by the National Polytechnique Institute (UPIITA-IPN) under multidisciplinary project SIP 20130440 and CONACyT-SEP project 183370.

Competing interests

The authors declare that they have no competing interests.

REFERENCES

1. Arampatzis T, Lygeros J, Manesis S: A survey of applications of wireless sensors and wireless sensor networks. In Proceedings of the 2005 IEEE International Symposium on Intelligent Control, 2005, Mediterranean Conference on Control and Automation. Limassol; June 2005:27-29.
2. Hu X, Wang B, Ji H: A wireless sensor network–based structural health monitoring system for highway bridges. Computer–Aided Civil and Infrastructure. Engineering 2012, 28(3):193-209.

3. Rivero-Angeles ME, Nizar B: Event reporting on continuous monitoring wireless sensor networks. In IEEE GlobeCom 2009. Honolulu; 2009. 30 November to 4 December
4. Heinzelman WR, Chandrakasan A, Balakrishnan H: Energy-efficient communication protocol for wireless microsensor networks. In Proceedings of the 33rd Annual Hawaii International Conference on System Sciences. Wailea Maui; 2000:4-7. January 2000
5. Paing TS, Zane R: Resistor emulation approach to low-power energy harvesting. In 37th IEEE Power Electronics Specialists Conference. PESC'06, Jeju; 2006:18-22. June 2006
6. Dwari S, Parsa L: Efficient low voltage direct AC/DC converters for self-powered wireless sensor nodes and mobile electronics. In IEEE 30th International Telecommunications Energy Conference, 2008. INTELEC 2008, San Diego; 2008:14-18. September 2008
7. Dayal R, Parsa L: A new single stage AC-DC converter for low voltage electromagnetic energy harvesting. In 2010 IEEE Energy Conversion Congress and Exposition (ECCE). Atlanta; 2010:12-16. September
8. Stutzman WL, Thiele GA: Chapter 6: wire antennas. In Antenna Theory and Design. Wiley, New York; 1998.
9. Marian V, Allard B, Vollaire C, Verdier J: Strategy for microwave energy harvesting from ambient field or a feeding source. IEEE Trans On Power Electronics 2012, 27(11):4481-4491.
10. Bouabdallah N, Rivero-Angeles ME, Sericola B: Continuous monitoring using event-driven reporting for cluster-based wireless sensor networks. IEEE Trans on Vehic. Tech. 2009, 58(7):3460-3479.
11. Rom R, Sidi M: Multiple Access Protocols. Performance and Analysis (Springer, Berlin; 1990. pp. 47–103

Chapter 3

The Eigen Theory of Electromagnetic Waves in Complex Media

Shaohua Guo

Zhejiang University of Science and Technology, P. R. China

1. INTRODUCTION

Since J. C. Maxwell presented the electromagnetic field equations in 1873, the existence of electromagnetic waves has been verified in various medium (Kong, 1986; Monk, 2003). But except for Helmholtz's equation of electromagnetic waves in isotropic media, the laws of propagation of electromagnetic waves in anisotropic media are not clear to us yet. For example, how many electromagnetic waves are there in anisotropic media? How fast can these electromagnetic waves propagate? Where are propagation direction and polarization direction of the electromagnetic waves? What are the space patterns of these waves? Although many research works were made in trying to deduce the equations of electromagnetic waves in anisotropic media based on the Maxwell's equation (Yakhno, 2005, 2006; Cohen, 2002; Haba, 2004), the explicit equations of electromagnetic waves in anisotropic media could not be obtained because the dielectric permittivity matrix and magnetic permeability matrix were all included in these equations, so that only local behaviour of electromagnetic waves, for example, in a certain plane or along a certain direction, can be studied.

On the other hand, it is a natural fact that electric and magnetic fields interact with each other in classical electromagnetics. Therefore, even if most of material studies deal with the properties due to dielectric polarisation, magneitc materials are also capable of producing quite interesting electro-magnetic effects (Lindellm et al., 1994). From the bi-anisotropic point of view, magnetic materials can be treated as a subclass of magnetoelectric materials. The linear constitutive relations linking the electric and magnetic fields to the electric and magnetic displacements contain four dyadics, three of which have direct magnetic contents. The magnetoelectric coupling has both theoretical and practical significance in solid state physics and materials science. Though first predicted by Pierre Curie, magnetoeletric coupling was originally through to be forbidden because it violates time-reversal symmetry, until Laudau and Lifshitz (Landau & Lifshitz, 1960) pointed out that time reversal is not a symmetry operation in some magnetic crystal. Based on this argument, Dzyaloshinskii (Dzyaloshinskii, 1959) predicted that magnetoelectric effect should occur in antiferromagnetic crystal Cr_2O_3, which was verified experimentally by Astrov (Astrov, 1960). Since then the magnetoelectric coupling has been observed in single-phase materials where simultaneous electric and magnetic ordering coexists, and in two-phase composites where the participating phase are pizoelectric and piezomagnetic (Bracke & Van Vliet,1981; Van Run et al., 1974)· Agyei and Birman (Agyei & Birman, 1990) carried out a detailed analysis of the linear magnetoelectric effect, which showed that the effect should occur not only in some magnetic but also in some electric crystals. Pradhan (Pradhan, 1993) showed that an electric charge placed in a magnetoelectric medium becomes a source of induced magnetic field with non-zero divergence of volume integral. Magnetoelectric effect in two-phase composites has been analyzed by Harshe et al. (Harshe et al., 1993), Nan (Nan, 1994) and Benveniste (Benveniste, 1995). Broadband transducers based on magnetoelectric effect have also been developed (Bracke & Van Vliet, 1981). Although the development mentioned above, no great progress in the theories of electromagnetic waves in bi-anisotropic media because of the difficulties in deal with the bi-coupling in electric field and magnetic one of the Maxell's equation and the bi-anisotropic constitutive equation by classical electromagnetic theory.

Recently there is a growing interest modeling and analysis of Maxwell's equations (Lee & Madsen, 1990; Monk, 1992; Jin et al., 1999). However, most work is restricted to simple medium such as air in the free space. On the other hand, we notice that lossy and dispersive media are ubiquitous, for example human tissue, water, soil, snow, ice, plasma, optical fibers and radar-absorbing materials. Hence the study of how electromagnetic wave interacts with dispersive media becomes very important. Some concrete applications include geophysical probing and subsurface studied of the moon and other planets (Bui et al., 1991), High power and ultra-wide-band radar systems, in which it is necessary to model ultra-wide-band electromagnetic pulse propagation through plasmas (Dvorak & Dudley, 1995), ground penetrating radar detection of buried objects in soil media (liu & Fan, 1999). The Debye medium plays an important role in electromagnetic wave interactions with biological and water-based substances (Gandhi & Furse, 1997). Until 1990, some paper on modeling of wave propagation in dispersive media started making their appearance in computational electromagnetics community. However, the published papers on modeling of dispersive media are exclusively restricted to the finite-difference time-domain methods and the finite element methods (Li & Chen, 2006; Lu et al., 2004). To our best knowledge, there exist only few works in the literature, which studied the theoretical model for the Maxwell's equation in the complex anisotropic dispersive media, and no explicit equations of electromagnetic waves in anisotropic dispersive media can be obtained due to the limitations of classical electromagnetic theory.

Chiral materials have been recently an interesting subject. In a chiral medium, an electric or magnetic excitation will produce simultaneously both electric and magnetic polarizations. On the other hand, the chiral medium is an object that cannot be brought into congruence with its mirror image by translation and rotation. Chirality is common in a variety of naturally occurring and man-made objects. From an operation point of view, chirality is introduced into the classical Maxwell equations by the Drude-Born-Fedorov relative constitutive relations in which the electric and magnetic fields are coupled via a new materials parameter (Lakhtakia, 1994; Lindell et al., 1994), the chirality parameter. These constitutive relations are chosen because they are symmetric under time reversality and duality transformations. In a homogeneous isotropic chiral medium the electromagnetic fields are composed of left-circularly polarized (LCP) and right- circularly polarized (RCP) components (Jaggard et al., 1979; Athanasiadis & Giotopoulos, 2003), which have different wave numbers and independent directions of propagation. Whenever an electromagnetic wave (LCP, RCP or a linear combination of them) is incident upon a chiral scatterer, then the scattered field is composed of both LCP and RCP components and therefore both LCP and RCP far-field patterns are derived. Hence, in the vector problem we need to specify two directions of propagation and two polarizations. In recent years, chiral materials have been increasingly studied and there is a growing literature covering both their applications and the theoretical investigation of their properties. It will be noticed that the works dealing with wave phenomena in chiral materials have been mainly concerned with the study of time-harmonic waves which lead to frequency domain studies (Lakhtakia et al., 1989; Athanasiadis et al., 2003).

In this chapter, the idea of standard spaces is used to deal with the Maxwell's electromagnetic equation (Guo, 2009, 2009, 2010, 2010, 2010). By this method, the classical Maxwell's equation under the geometric presentation can be transformed into the eigen Maxwell's equation under the physical presentation. The former is in the form of vector and the latter is in the form of scalar. Through inducing the modal constitutive equations of complex media, such as anisotropic media, bi-anisotropic media, lossy media, dissipative media, and chiral media, a set of modal equations of electromagnetic waves for all of those media are obtained, each of which shows the existence of electromagnetic sub-waves, meanwhile its propagation velocity, propagation direction, polarization direction and space pattern can be completely determined by the modal equations.This chapter will make introductions of the eigen theory to reader in details. Several novel theoretical results were discussed in the different parts of this chapter.

2. STANDARD SPACES OF ELECTROMAGNETIC MEDIA

In anisotropic electromagnetic media, the dielectric permittivity and magnetic permeability are tensors instead of scalars. The constitutive relations are expressed as follows

$$\boldsymbol{D} = \boldsymbol{\varepsilon} \cdot \boldsymbol{E}, \quad \boldsymbol{B} = \boldsymbol{\mu} \cdot \boldsymbol{H} \tag{1}$$

Rewriting Eq.(1) in form of scalar, we have

$$D_i = \varepsilon_{ij} E_j, \quad B_i = \mu_{ij} H_j \tag{2}$$

where the dielectric permittivity matrix $\varepsilon$and the magnetic permeability matrix μ are usually symmetric ones, and the elements of the matrixes have a close relationship with the selection of reference coordinate. Suppose that if the reference coordinates is selected along principal axis of electrically or magnetically anisotropic media, the elements at non-diagonal of these matrixes turn to be zero. Therefore, equations (1) and (2) are called the constitutive equations of electromagnetic media under the geometric presentation. Now we intend to get rid of effects of geometric coordinate on the constitutive equations, and establish a set of coordinate-independent constitutive equations of electromagnetic media under physical presentation. For this purpose, we solve the following problems of eigen-value of matrixes.

$$(\boldsymbol{\varepsilon} - \lambda \boldsymbol{I})\boldsymbol{\phi} = \boldsymbol{0}, \quad (\boldsymbol{\mu} - \gamma \boldsymbol{I})\boldsymbol{\varphi} = \boldsymbol{0} \tag{3}$$

where λi(i=1,2,3)and γi(i=1,2,3)are respectively eigen dielectric permittivity and eigen magnetic permeability, which are constants of coordinate-independent. ϕi(i=1,2,3) and φi(i=1,2,3) are respectively eigen electric vector and eigen magnetic vector, which show the electrically principal direction and magnetically principal direction of anisotropic media, and are all coordinate-dependent. We call these vectors as standard spaces. Thus, the matrix of dielectric permittivity and magnetic permeability can be spectrally decomposed as follows

$$\varepsilon = \Phi \Lambda \Phi^{\mathrm{T}}, \quad \mu = \Psi \Pi \Psi^{\mathrm{T}} \tag{4}$$

where $\Lambda = diag[\lambda 1, \lambda 2, \lambda 3]$ and $\Pi = diag[\gamma 1, \gamma 2, \gamma 3]$ are the matrix of eigen dielectric permittivity and eigen magnetic permeability, respectively. $\Phi = \{\phi 1, \phi 2, \phi 3\}$ and $\Psi = \{\varphi 1, \varphi 2, \varphi 3\}$ are respectively the modal matrix of electric media and magnetic media, which are both orthogonal and positive definite matrixes, and satisfy $\Phi T \Phi = I$, $\Psi T \Psi = I$ Projecting the electromagnetic physical qualities of the geometric presentation, such as the electric field intensity vector *E*, magnetic field intensity vector *H*, magnetic flux density vector *B*and electric displacement vector *D*into the standard spaces of the physical presentation, we get

$$\boldsymbol{D}^{*} = \boldsymbol{\Phi}^{\mathrm{T}} \cdot \boldsymbol{D}, \quad \boldsymbol{E}^{*} = \boldsymbol{\Phi}^{\mathrm{T}} \cdot \boldsymbol{E} \tag{5}$$

$$\boldsymbol{B}^{*} = \boldsymbol{\Psi}^{\mathrm{T}} \cdot \boldsymbol{B}, \quad \boldsymbol{H}^{*} = \boldsymbol{\Psi}^{\mathrm{T}} \cdot \boldsymbol{H} \tag{6}$$

Rewriting Eqs.(5) and (6) in the form of scalar, we have

$$D_i^{*} = \boldsymbol{\phi}_i^{\mathrm{T}} \cdot \boldsymbol{D} \quad i = 1,2,3, \quad E_i^{*} = \boldsymbol{\phi}_i^{\mathrm{T}} \cdot \boldsymbol{E} \quad i = 1,2,3 \tag{7}$$

$$B_i^{*} = \boldsymbol{\varphi}_i^{\mathrm{T}} \cdot \boldsymbol{B} \quad i = 1,2,3, \quad H_i^{*} = \boldsymbol{\varphi}_i^{\mathrm{T}} \cdot \boldsymbol{H} \quad i = 1,2,3 \tag{8}$$

These are the electromagnetic physical qualities under the physical presentation.

Substituting Eq. (4) into Eq. (1) respectively, and using Eqs.(5) and (6) yield

$$D_i^{*} = \lambda_i E_i^{*} \quad i = 1,2,3 \tag{9}$$

$$B_i^{*} = \gamma_i H_i^{*} \quad i = 1,2,3 \tag{10}$$

The above equations are just the modal constitutive equations in the form of scalar.

3. EIGEN EXPRESSION OF MAXWELL'S EQUATION

The classical Maxwell's equations in passive region can be written as

$$\nabla \times \boldsymbol{H} = \nabla_t \boldsymbol{D} \quad , \quad \nabla \times \boldsymbol{E} = -\nabla_t \boldsymbol{B} \tag{11}$$

Now we rewrite the equations in the form of matrix as follows

$$\begin{bmatrix} 0 & -\partial_z & \partial_y \\ \partial_z & 0 & -\partial_x \\ -\partial_y & \partial_x & 0 \end{bmatrix} \begin{Bmatrix} H_1 \\ H_2 \\ H_3 \end{Bmatrix} = \nabla_t \begin{Bmatrix} D_1 \\ D_2 \\ D_3 \end{Bmatrix} \tag{12}$$

or

$$[\Delta]\{H\} = \nabla_t \{D\} \tag{13}$$

$$\begin{bmatrix} 0 & -\partial_z & \partial_y \\ \partial_z & 0 & -\partial_x \\ -\partial_y & \partial_x & 0 \end{bmatrix} \begin{Bmatrix} E_1 \\ E_2 \\ E_3 \end{Bmatrix} = -\nabla_t \begin{Bmatrix} B_1 \\ B_2 \\ B_3 \end{Bmatrix} \tag{14}$$

or

$$[\Delta]\{E\} = -\nabla_t \{B\} \tag{15}$$

where $[\Delta]$is defined as the matrix of electric and magnetic operators.Substituting Eq. (1) into Eqs. (13) and (15) respectively, we have

$$[\Delta]\{H\} = \nabla_t [\varepsilon]\{E\} \tag{16}$$

$$[\Delta]\{E\} = -\nabla_t [\mu]\{H\} \tag{17}$$

Substituting Eq. (16) into (17) or Eq. (17) into (16), yield

$$[\Box]\{H\}=-\nabla_t^2[\mu][\varepsilon]\{H\} \tag{18}$$

$$[\Box]\{E\}=-\nabla_t^2[\mu][\varepsilon]\{E\} \tag{19}$$

where $[\Box]=[\Delta][\Delta]$is defined as the matrix of electromagnetic operators as follows

$$[\Box]=\begin{bmatrix} -(\partial_z^2+\partial_y^2) & \partial_{xy}^2 & \partial_{xz}^2 \\ \partial_{yx}^2 & -(\partial_x^2+\partial_z^2) & \partial_{yz}^2 \\ \partial_{zx}^2 & \partial_{zy}^2 & -(\partial_x^2+\partial_y^2) \end{bmatrix} \tag{20}$$

In another way, substituting Eqs. (5) and (6) into Eqs. (13) and (15), respectively, we have

$$[\Delta][\Phi]\{E^*\}=-\nabla_t[\Psi]\{B^*\} \tag{21}$$

$$[\Delta][\Psi]\{H^*\}=\nabla_t[\Phi]\{D^*\} \tag{22}$$

Rewriting the above in indicial notation, we get

$$\{\Delta_i^*\}E_i^*=-\nabla_t\{\varphi_i\}B_i^* \qquad i=1,2,3 \tag{23}$$

$$\{\Delta_i^*\}H_i^*=\nabla_t\{\phi_i\}D_i^* \qquad i=1,2,3 \tag{24}$$

where, $\Delta*I$ is the electromagnetic intensity operator, and I th row of

$$[\Delta^*]=[\Delta][\Phi]$$

4. ELECTROMAGNETIC WAVES IN ANISOTROPIC MEDIA

4.1. Electrically anisotropic media

In anisotropic dielectrics, the dielectric permittivity is a tensor, while the magnetic permeability is a scalar. So Eqs. (18) and (19) can be written as follows

$$[\Box]\{H\}=-\nabla_t^2\mu_0[\varepsilon]\{H\} \tag{25}$$

$$[\Box]\{E\} = -\nabla_t^2 \mu_0 [\varepsilon]\{E\} \tag{26}$$

Substituting Eqs. (4) - (6) into Eqs. (25) and (26), we have

$$[\Box^*]\{H^*\} = -\nabla_t^2 \mu_0 [\Lambda]\{H^*\} \tag{27}$$

$$[\Box^*]\{E^*\} = -\nabla_t^2 \mu_0 [\Lambda]\{E^*\} \tag{28}$$

where $[\Box*]=[\Phi]T[\Box][\Phi]$is defined as the eigen matrix of electromagnetic operators under the standard spaces. We can note from Appendix A that it is a diagonal matrix. Thus Eqs. (27) and (28) can be uncoupled in the form of scalar

$$\Box_i^* H_i^* + \mu_0 \lambda_i \nabla_t^2 H_i^* = 0 \quad i = 1,2,3 \tag{29}$$

$$\Box_i^* E_i^* + \mu_0 \lambda_i \nabla_t^2 E_i^* = 0 \quad i = 1,2,3 \tag{30}$$

Eqs.(29) and (30) are the modal equations of electromagnetic waves in anisotropic dielectrics.

4.2. Magnetically anisotropic media

In anisotropic magnetics, the magnetic permeability is a tensor, while the dielectric permittivity is a scalar. So Eqs. (18) and (19) can be written as follows

$$[\Box]\{H\} = -\nabla_t^2 \varepsilon_0 [\mu]\{H\} \tag{31}$$

$$[\Box]\{E\} = -\nabla_t^2 \varepsilon_0 [\mu]\{E\} \tag{32}$$

Substituting Eqs. (4) - (6) into Eqs. (31) and (32), we have

$$[\Box^*]\{H^*\} = -\nabla_t^2 \varepsilon_0 [\Pi]\{H^*\} \tag{33}$$

$$[\Box^*]\{E^*\} = -\nabla_t^2 \varepsilon_0 [\Pi]\{E^*\} \tag{34}$$

where [□∗]=[Ψ]T[□][Ψ]is defined as the eigen matrix of electromagnetic operators under the standard spaces. We can also note from Appendix A that it is a diagonal matrix. Thus Eqs. (33) and (34) can be uncoupled in the form of scalar

$$\square_i^* H_i^* + \varepsilon_0 \gamma_i \nabla_t^2 H_i^* = 0 \qquad i = 1,2,3 \tag{35}$$

$$\square_i^* E_i^* + \varepsilon_0 \gamma_i \nabla_t^2 E_i^* = 0 \qquad i = 1,2,3 \tag{36}$$

Eqs.(35) and (36) are the modal equations of electromagnetic waves in anisotropic magnetics.

5. ELECTROMAGNETIC WAVES IN BI-ANISOTROPIC MEDIA

5.1. Bi-anisotropic constitutive equations

The constitutive equations of bi-anisotropic media are the following (Lindellm & Sihvola, 1994; Laudau & Lifshitz, 1960)

$$\boldsymbol{D} = \boldsymbol{\varepsilon} \cdot \boldsymbol{E} + \boldsymbol{\xi} \cdot \boldsymbol{H} \tag{37}$$

$$\boldsymbol{B} = \boldsymbol{\xi} \cdot \boldsymbol{E} + \boldsymbol{\mu} \cdot \boldsymbol{H} \tag{38}$$

where ξ is the matrix of magneto-electric parameter, and a symmetric one. Substituting Eqs. (5) and (6) into Eqs. (37) and (38), respectively, and multiplying them with the transpose of modal matrix in the left, we have

$$\boldsymbol{\Phi}^{\mathrm{T}} \boldsymbol{D} = \boldsymbol{\Phi}^{\mathrm{T}} \boldsymbol{\varepsilon} \boldsymbol{\Phi} \boldsymbol{E}^* + \boldsymbol{\Phi}^{\mathrm{T}} \boldsymbol{\xi} \boldsymbol{\Psi} \boldsymbol{H}^* \tag{39}$$

$$\boldsymbol{\Psi}^{\mathrm{T}} \boldsymbol{B} = \boldsymbol{\Psi}^{\mathrm{T}} \boldsymbol{\xi} \boldsymbol{\Phi} \boldsymbol{E}^* + \boldsymbol{\Psi}^{\mathrm{T}} \boldsymbol{\mu} \boldsymbol{\Psi} \boldsymbol{H}^* \tag{40}$$

Let *G=ΦTξΨ=ΨTξΦ*, that is a coupled magneto-electric matrix, and using Eq. (4), we have

$$\boldsymbol{D}^* = \boldsymbol{\Lambda} \boldsymbol{E}^* + \boldsymbol{G} \boldsymbol{H}^* \tag{41}$$

$$B^* = GE^* + \Pi H^* \tag{42}$$

Rewriting the above in indicial notation, we get

$$D_i^* = \lambda_i E_i^* + g_{ii} H_i^* \quad i = 1,2,3 \quad j = 1,2,3 \tag{43}$$

$$B_i^* = \gamma_i H_i^* + g_{ij} E_j^* \quad i = 1,2,3 \quad j = 1,2,3 \tag{44}$$

Eqs. (43) and (44) are just the modal constitutive equations for bi-anisotropic media.

5.2. Eigen equations of electromagnetic waves in bi-anisotropic media

Substituting Eqs. (43) and (44) into Eqs. (23) and (24), respectively, we have

$$\{\Delta_i^*\} E_i^* = -\nabla_t \{\varphi_i\} \left(\gamma_i H_i^* + g_{ij} E_j^* \right) \tag{45}$$

$$\{\Delta_i^*\} H_i^* = \nabla_t \{\phi_i\} \left(\lambda_i E_i^* + g_{ij} H_j^* \right) \tag{46}$$

From them, we can get

$$\left(\{\Lambda_i^*\} - \nabla_t \{\phi_i\} g_{ij} \delta_{ij} \right)^T \left(\{\Delta_i^*\} + \nabla_t \{\varphi_i\} g_{ij} \delta_{ij} \right) E_i^* = -\nabla_t^2 \{\varphi_i\} \{\phi_i\}^T \lambda_i \gamma_i E_i^* \tag{47}$$

$$\left(\{\Delta_i^*\} + \nabla_t \{\varphi_i\} g_{ij} \delta_{ij} \right)^T \left(\{\Delta_i^*\} - \nabla_t \{\phi_i\} g_{ij} \delta_{ij} \right) H_i^* = -\nabla_t^2 \{\phi_i\} \{\varphi_i\}^T \gamma_i \lambda_i H_i^* \tag{48}$$

The above can also be written as the standard form of waves

$$\square_i^* E_i^* + \nabla_t \{\Delta_i^*\}^T \cdot (\{\varphi\} - \{\phi\}) g_{ii} E_i^* + \nabla_t^2 \{\phi\}^T \cdot \{\varphi\} (\lambda_i \gamma_i - g_{ii}^2) E_i^* = 0 \quad i = 1,2,3 \tag{49}$$

$$\square_i^* H_i^* + \nabla_t \{\Delta_i^*\}^T \cdot (\{\varphi\} - \{\phi\}) g_{ii} H_i^* + \nabla_t^2 \{\phi\}^T \cdot \{\varphi\} (\lambda_i \gamma_i - g_{ii}^2) H_i^* = 0 \quad i = 1,2,3 \tag{50}$$

where, □∗i={Δ∗i}T·{Δ∗i}is the electromagnetic operator. Eqs.(49) and (50) are just equations of electric field and magnetic field for bi-anisotropic media.

5.3. Applications

5.3.1. Bi-isotropic media

The constitutive equations of bi-isotropic media are the following

$$\boldsymbol{D} = \begin{bmatrix} \varepsilon & 0 & 0 \\ 0 & \varepsilon & 0 \\ 0 & 0 & \varepsilon \end{bmatrix} \cdot \boldsymbol{E} + \begin{bmatrix} \xi & 0 & 0 \\ 0 & \xi & 0 \\ 0 & 0 & \xi \end{bmatrix} \cdot \boldsymbol{H} \tag{51}$$

$$\boldsymbol{B} = \begin{bmatrix} \xi & 0 & 0 \\ 0 & \xi & 0 \\ 0 & 0 & \xi \end{bmatrix} \cdot \boldsymbol{E} + \begin{bmatrix} \mu & 0 & 0 \\ 0 & \mu & 0 \\ 0 & 0 & \mu \end{bmatrix} \cdot \boldsymbol{H} \tag{52}$$

The eigen values and eigen vectors of those matrix are the following

$$\boldsymbol{\Lambda} = diag[\varepsilon, \varepsilon, \varepsilon] \ , \quad \boldsymbol{\Pi} = diag[\mu, \mu, \mu] \tag{53}$$

$$\boldsymbol{\Phi} = \boldsymbol{\Psi} = \begin{bmatrix} 1 & 0 & 0 \\ 0 & 1 & 0 \\ 0 & 0 & 1 \end{bmatrix} \tag{54}$$

We can see from the above equations that there is only one eigen-space in isotropic medium, which is a triple-degenerate one, and the space structure is the following

$$\boldsymbol{W} = W_1^{(3)} [\phi_1, \phi_2, \phi_3] \tag{55}$$

$$\phi_1^* = \varphi_1^* = \frac{\sqrt{3}}{3} [1,1,1]^T \tag{56}$$

Then the eigen-qualities and eigen-operators of bi-isotropic medium are respectively shown as belows

$$E_1^* = \phi_1^{*\mathrm{T}} \cdot \boldsymbol{E} = \frac{\sqrt{3}}{3} (E_1 + E_2 + E_3) \tag{57}$$

$$\square_1^* = -\left(\partial_x^2 + \partial_y^2 + \partial_z^2\right) , \ g_{11} = \xi \tag{58}$$

So, the equation of electromagnetic wave in bi-isotropic medium becomes

$$\left(\partial_x^2+\partial_y^2+\partial_z^2\right)E_1^* = \left(\mu\varepsilon-\xi^2\right)\partial_t^2 E_1^* \tag{59}$$

the velocity of electromagnetic wave is

$$c^{(1)} = \frac{1}{\sqrt{\mu\varepsilon-\xi^2}} \tag{60}$$

5.3.2. Dzyaloshinskii's bi-anisotropic media

Dzyaloshinskii's constitutive equations of bi-anisotropic media are the following

$$\boldsymbol{D} = \begin{bmatrix} \varepsilon & 0 & 0 \\ 0 & \varepsilon & 0 \\ 0 & 0 & \varepsilon_z \end{bmatrix} \cdot \boldsymbol{E} + \begin{bmatrix} \xi & 0 & 0 \\ 0 & \xi & 0 \\ 0 & 0 & \xi_z \end{bmatrix} \cdot \boldsymbol{H} \tag{61}$$

$$\boldsymbol{B} = \begin{bmatrix} \xi & 0 & 0 \\ 0 & \xi & 0 \\ 0 & 0 & \xi_z \end{bmatrix} \cdot \boldsymbol{E} + \begin{bmatrix} \mu & 0 & 0 \\ 0 & \mu & 0 \\ 0 & 0 & \mu_z \end{bmatrix} \cdot \boldsymbol{H} \tag{62}$$

The eigen values and eigen vectors of those matrix are the following

$$\boldsymbol{\Lambda} = diag\left[\varepsilon,\varepsilon,\varepsilon_z\right], \quad \boldsymbol{\Pi} = diag\left[\mu,\mu,\mu_z\right] \tag{63}$$

$$\boldsymbol{\Phi} = \boldsymbol{\Psi} = \begin{bmatrix} 1 & 0 & 0 \\ 0 & 1 & 0 \\ 0 & 0 & 1 \end{bmatrix} \tag{64}$$

We can see from the above equations that there are two eigen-spaces in Dzyaloshinskii's bi-anisotropic medium, in which one is a binary-degenerate one, the space structure is the following

$$\boldsymbol{W} = W_1^{(2)}\left[\phi_1,\phi_2\right] \oplus W_2^1\left[\phi_3\right] \tag{65}$$

Then the eigen-qualities and eigen-operators of Dzyaloshinskii's bi-anisotropic medium are respectively shown as belows

$$E_2^* = \phi_2^{\mathrm{T}} \cdot \boldsymbol{E} = E_3\ ,\quad \left|E_1^*\right| = \sqrt{\left(\boldsymbol{E} - \phi_2^{\mathrm{T}} E_2^*\right)^{\mathrm{T}} \left(\boldsymbol{E} - \phi_2^{\mathrm{T}} E_2^*\right)} = \sqrt{E_1^2 + E_2^2} \tag{66}$$

$$\square_1^* = -\left(\partial_x^2 + \partial_y^2 + 2\partial_z^2 - 2\partial_{xy}^2\right),\ \square_2^* = -\left(\partial_x^2 + \partial_y^2\right),\ g_{11} = \xi,\ g_{22} = \xi_z \tag{67}$$

So, the equations of electromagnetic wave in Dzyaloshinskii's bi-anisotropic medium become

$$\left(\partial_x^2 + \partial_y^2 + 2\partial_z^2 - 2\partial_{xy}^2\right)\sqrt{E_1^2 + E_2^2} = \left(\mu\varepsilon - \xi^2\right)\partial_t^2 \sqrt{E_1^2 + E_2^2} \tag{68}$$

$$\left(\partial_x^2 + \partial_y^2\right)E_3 = \left(\mu_z\varepsilon_z - \xi_z^2\right)\partial_t^2 E_3 \tag{69}$$

the velocities of electromagnetic wave are

$$c^{(1)} = \frac{1}{\sqrt{\mu\varepsilon - \xi^2}} \tag{70}$$

$$c^{(2)} = \frac{1}{\sqrt{\mu_z\varepsilon_z - \xi_z^2}} \tag{71}$$

It is seen both from bi-isotropic media and Dzyaloshinskii's bi-anisotropic medium that the electromagnetic waves in bi-anisotropic medium will go faster duo to the bi-coupling between electric field and magnetic one.

6. ELECTROMAGNETIC WAVES IN LOSSY MEDIA

6.1. The constitutive equation of lossy media

The constitutive equation of lossy media is the following

$$\boldsymbol{D} = \boldsymbol{\varepsilon} \cdot \boldsymbol{E} + \int_t \sigma \cdot \frac{d\boldsymbol{E}}{d\tau} d\tau \tag{72}$$

It is equivalent to the following differential constitutive equation

$$\dot{\boldsymbol{D}} = \boldsymbol{\varepsilon} \cdot \dot{\boldsymbol{E}} + \boldsymbol{\sigma} \cdot \boldsymbol{E} \tag{73}$$

Let

$$\boldsymbol{D}^{\mathrm{e}} = \boldsymbol{\varepsilon} \cdot \boldsymbol{E} \ , \ \dot{\boldsymbol{D}}^{\mathrm{d}} = \boldsymbol{\sigma} \cdot \boldsymbol{E} \tag{74}$$

Eq.(73) can be written as

$$\dot{\boldsymbol{D}} = \dot{\boldsymbol{D}}^{\mathrm{e}} + \dot{\boldsymbol{D}}^{\mathrm{d}} \tag{75}$$

or

$$\nabla_t \{D\} = ([\varepsilon]\nabla_t + [\sigma])\{E\} \tag{76}$$

Using Eq.(5), the above becomes

$$\nabla_t \{D^*\} = \left([\Phi]^T [\varepsilon][\Phi]\nabla_t + [\Phi]^T [\sigma][\Phi]\right)\{E^*\} \tag{77}$$

According to Appendix B and Eq.(77), we have

$$\nabla_t \{D^*\} = ([\Lambda]\nabla_t + [\Gamma])\{E^*\} \tag{78}$$

Rewriting the above in indicial notation, we get

$$\nabla_t D_i^* = (\lambda_i \nabla_t + \eta_i) E_i^* \tag{79}$$

Eq.(79) is just the modal constitutive equations for lossy media.

6.2. Eigen equations of electromagnetic waves in lossy media

Substituting Eqs. (10) and (79) into Eqs. (23) and (24), respectively, we have

$$\{\Delta_i^*\} E_i^* = -\nabla_t \{\varphi_i\} \gamma_i H_i^* \quad i = 1,2,3 \tag{80}$$

$$\{\Delta_i^*\} H_i^* = \{\phi_i\} (\lambda_i \nabla_t + \eta_i) E_i^* \quad i = 1,2,3 \tag{81}$$

From them, we can get

$$\overset{*}{\square}_i E_i^* + \nabla_{tt}\xi_i\gamma_i\lambda_i E_i^* + \nabla_t\xi_i\gamma_i\eta_i E_i^* = 0 \quad i=1,2,3 \tag{82}$$

$$\overset{*}{\square}_i H_i^* + \nabla_{tt}\xi_i\gamma_i\lambda_i H_i^* + \nabla_t\xi_i\gamma_i\eta_i H_i^* = 0 \quad i=1,2,3 \tag{83}$$

where $\xi i=\{\phi*i\}T\cdot\{\varphi*i\}$. Eqs.(82) and (83) are just equations of electric field and magnetic field for bi-anisotropic media.

6.3. Applications

In this section, we discuss the propagation laws of electromagnetic waves in an isotropic lossy medium. The material tensors in Eqs.(1) and (72) are represented by the following matrices

$$\boldsymbol{\varepsilon} = \begin{bmatrix} \varepsilon_{11} & 0 & 0 \\ 0 & \varepsilon_{11} & 0 \\ 0 & 0 & \varepsilon_{11} \end{bmatrix}, \boldsymbol{\mu} = \begin{bmatrix} \mu_{11} & 0 & 0 \\ 0 & \mu_{11} & 0 \\ 0 & 0 & \mu_{11} \end{bmatrix}, \boldsymbol{\sigma} = \begin{bmatrix} \sigma_{11} & 0 & 0 \\ 0 & \sigma_{11} & 0 \\ 0 & 0 & \sigma_{11} \end{bmatrix} \tag{84}$$

The eigen values and eigen vectors of those matrix are the following

$$\boldsymbol{\Lambda} = diag[\varepsilon_{11}, \varepsilon_{11}, \varepsilon_{11}],\ \boldsymbol{\Pi} = diag[\mu_{11}, \mu_{11}, \mu_{11}],\ \boldsymbol{\Gamma} = diag[\sigma_{11}, \sigma_{11}, \sigma_{11}] \tag{85}$$

$$\boldsymbol{\Phi} = \boldsymbol{\Psi} = \boldsymbol{\Theta} = \begin{bmatrix} 1 & 0 & 0 \\ 0 & 1 & 0 \\ 0 & 0 & 1 \end{bmatrix} \tag{86}$$

We can see from the above equations that there is only one eigen-space in an isotropic lossy medium, which is a triple-degenerate one, and the space structure is the following.

$$\boldsymbol{W}_{mag} = W_1^{(3)}[\phi_1, \phi_2, \phi_3],\ \boldsymbol{W}_{ele} = W_1^{(3)}[\varphi_1, \varphi_2, \varphi_3] \tag{87}$$

where, $\phi_1^* = \frac{\sqrt{3}}{3}\{1,1,1\}^T$, $\varphi_1^* = \frac{\sqrt{3}}{3}\{1,1,1\}^T$, $\xi_1 = 1$.

Then the eigen-qualities and eigen-operators of an isotropic lossy media are respectively shown as follows

$$E_1^* = \frac{\sqrt{3}}{3}(E_1 + E_2 + E_3) \tag{88}$$

$$H_1^* = \frac{\sqrt{3}}{3}(H_1 + H_2 + H_3) \tag{89}$$

$$\square_1^* = \frac{1}{3}\left[-\left(\partial_x^2 + \partial_y^2 + \partial_z^2\right)\right] \tag{90}$$

So, the equation of electromagnetic wave in lossy media becomes

$$\left(\partial_x^2 + \partial_y^2 + \partial_z^2\right)E_1^* = \frac{1}{c^2}\partial_t^2 E_1^* + \frac{1}{\tau^2}\partial_t E_1^* \tag{91}$$

Rewriting it in the component form, we have

$$\left(\partial_x^2 + \partial_y^2 + \partial_z^2\right)E_1 = \frac{1}{c^2}\partial_t^2 E_1 + \frac{1}{\tau^2}\partial_t E_1 \tag{92}$$

$$\left(\partial_x^2 + \partial_y^2 + \partial_z^2\right)E_2 = \frac{1}{c^2}\partial_t^2 E_2 + \frac{1}{\tau^2}\partial_t E_2 \tag{93}$$

$$\left(\partial_x^2 + \partial_y^2 + \partial_z^2\right)E_3 = \frac{1}{c^2}\partial_t^2 E_3 + \frac{1}{\tau^2}\partial_t E_3 \tag{94}$$

where, $c$is the velocity of electromagnetic wave, $\tau$is the lossy coefficient of electromagnetic wave

$$c = \frac{1}{\sqrt{\mu_{11}\varepsilon_{11}}}, \quad \tau = \frac{1}{\sqrt{\mu_{11}\sigma_{11}}} \tag{95}$$

Now, we discuss the the propagation laws of a plane electromagnetic wave in x-axis. In this time, Eq. (92) becomes

$$\frac{\partial}{\partial x^2}E_1 = \frac{1}{c^2}\partial_t^2 E_1 + \frac{1}{\tau^2}\partial_t E_1 \tag{96}$$

Let the solution of Eq. (96) is as follows

$$E_1 = A\exp\left[i(kx-\omega t)\right] \tag{97}$$

Substituting the above into Eq. (96), we have

$$k^2 = \frac{\omega^2}{c^2} + i\frac{\omega}{\tau^2} \tag{98}$$

From Eq.(96), we can get

$$\overline{k} = k_1 + ik_2 \tag{99}$$

Where

$$k_1 = \frac{\omega}{c}\left[\frac{1+\left(1+\dfrac{c^4}{\omega^2\tau^4}\right)^{\frac{1}{2}}}{2}\right]^{\frac{1}{2}}, \quad k_2 = \frac{\omega}{c}\left[\frac{-1+\left(1+\dfrac{c^4}{\omega^2\tau^4}\right)^{\frac{1}{2}}}{2}\right]^{\frac{1}{2}}.$$

Then, the solutions of electromagnetic waves are the following

$$E_1 = Ae^{-k_2 x}\cdot e^{i(k_1 x-\omega t)} = \overline{A}\cdot e^{i(k_1 x-\omega t)} \tag{100}$$

It is an attenuated sub-waves.

7. ELECTROMAGNETIC WAVES IN DISPERSIVE MEDIA

7.1. The constitutive equation of dispersive media

The general constitutive equations of dispersive media are the following

$$\boldsymbol{D} = \boldsymbol{\varepsilon}\cdot\boldsymbol{E} + \boldsymbol{\varepsilon}_1\cdot\dot{\boldsymbol{E}} + \boldsymbol{\varepsilon}_2\cdot\ddot{\boldsymbol{E}} + \cdots \tag{101}$$

$$\boldsymbol{B} = \boldsymbol{\mu}\cdot\boldsymbol{H} + \boldsymbol{\mu}_1\cdot\dot{\boldsymbol{H}} + \boldsymbol{\mu}_2\cdot\ddot{\boldsymbol{H}} + \cdots \tag{102}$$

where $\varepsilon i,(i=1,2,\cdots)$and $\mu i,(i=1,2,\cdots)$are the higher order dielectric permittivity matrix and the magnetic permeability matrix respectively, and all symmtric ones. Substituting Eqs. (5) and (6) into Eqs. (101) and

(102), respectively, and multiplying them with the transpose of modal matrix in the left, we have

$$\boldsymbol{D}^{*}=\boldsymbol{\Phi}^{T}\boldsymbol{\varepsilon}\boldsymbol{\Phi}\boldsymbol{E}^{*}+\boldsymbol{\Phi}^{T}\boldsymbol{\varepsilon}_{1}\boldsymbol{\Phi}\dot{\boldsymbol{E}}+\boldsymbol{\Phi}^{T}\boldsymbol{\varepsilon}_{2}\boldsymbol{\Phi}\ddot{\boldsymbol{E}}+\cdots \tag{103}$$

$$\boldsymbol{B}^{*}=\boldsymbol{\Psi}^{T}\boldsymbol{\mu}\boldsymbol{\Psi}\boldsymbol{H}+\boldsymbol{\Psi}^{T}\boldsymbol{\mu}_{1}\boldsymbol{\Psi}\dot{\boldsymbol{H}}+\boldsymbol{\Psi}^{T}\boldsymbol{\mu}_{2}\boldsymbol{\Psi}\ddot{\boldsymbol{H}}+\cdots \tag{104}$$

It can be proved that there exist same standard spaces for various order electric and magnetic fields in the condition close to the thermodynamic equilibrium. Then, we have

$$D_{i}^{*}=\lambda_{i}E_{i}^{*}+\lambda_{i}^{(1)}\dot{E}_{i}^{*}+\lambda_{i}^{(2)}\ddot{E}_{i}^{*}+\cdots \tag{105}$$

$$B_{i}^{*}=\gamma_{i}H_{i}^{*}+\gamma_{i}^{(1)}\dot{H}_{i}^{*}+\gamma_{i}^{(2)}\ddot{H}_{i}^{*}+\cdots \tag{106}$$

Eqs. (105) and (106) are just the modal constitutive equations for the general dispersive media.

7.2. Eigen equations of electromagnetic waves in dispersive media

Substituting Eqs. (105) and (106) into Eqs. (23) and (24), respectively, we have

$$\left\{\Delta_{i}^{*}\right\}E_{i}^{*}=-\nabla_{t}\left\{\varphi_{i}\right\}\left(\gamma_{i}H_{i}^{*}+\gamma_{i}^{(1)}\dot{H}_{i}^{*}+\gamma_{i}^{(2)}\ddot{H}_{i}^{*}+\cdots\right) \tag{107}$$

$$\left\{\Delta_{i}^{*}\right\}H_{i}^{*}=\nabla_{t}\left\{\phi_{i}\right\}\left(\lambda_{i}E_{i}^{*}+\lambda_{i}^{(1)}\dot{E}_{i}^{*}+\lambda_{i}^{(2)}\ddot{E}_{i}^{*}+\cdots\right) \tag{108}$$

From them, we can get

$$\frac{1}{\xi_{i}}\Box_{i}^{*}E_{i}^{*}+\gamma_{i}\lambda_{i}\nabla_{tt}E_{i}^{*}+\left(\gamma_{i}\lambda_{i}^{(1)}+\gamma_{i}^{(1)}\lambda_{i}\right)\nabla_{ttt}E_{i}^{*}+\left(\gamma_{i}\lambda_{i}^{(2)}+\gamma_{i}^{(1)}\lambda_{i}^{(1)}+\gamma_{i}^{(2)}\lambda_{i}\right)\nabla_{tttt}E_{i}^{*}+\cdots=0 \tag{109}$$

$$\frac{1}{\xi_{i}}\Box_{i}^{*}H_{i}^{*}+\gamma_{i}\lambda_{i}\nabla_{tt}H_{i}^{*}+\left(\gamma_{i}\lambda_{i}^{(1)}+\gamma_{i}^{(1)}\lambda_{i}\right)\nabla_{ttt}H_{i}^{*}+\left(\gamma_{i}\lambda_{i}^{(2)}+\gamma_{i}^{(1)}\lambda_{i}^{(1)}+\gamma_{i}^{(2)}\lambda_{i}\right)\nabla_{tttt}H_{i}^{*}+\cdots=0 \tag{110}$$

Eqs.(109) and (110) are just equations of electric field and magnetic field for general dispersive media.

7.3. Applications

In this section, we discuss the propagation laws of electromagnetic waves in an one-order dispersive medium. The material tensors in Eqs.(101) and (102) are represented by the following matrices

$$\boldsymbol{\varepsilon}=\begin{bmatrix}\varepsilon_{11} & 0 & 0\\ 0 & \varepsilon_{11} & 0\\ 0 & 0 & \varepsilon_{11}\end{bmatrix},\ \boldsymbol{\mu}=\begin{bmatrix}\mu_{11} & 0 & 0\\ 0 & \mu_{11} & 0\\ 0 & 0 & \mu_{11}\end{bmatrix},\ \boldsymbol{\varepsilon}_1=\begin{bmatrix}\varepsilon'_{11} & 0 & 0\\ 0 & \varepsilon'_{11} & 0\\ 0 & 0 & \varepsilon'_{11}\end{bmatrix},\ \boldsymbol{\mu}_1=\begin{bmatrix}\mu'_{11} & 0 & 0\\ 0 & \mu'_{11} & 0\\ 0 & 0 & \mu'_{11}\end{bmatrix} \tag{111}$$

The eigen values and eigen vectors of those matrix are the following

$$\boldsymbol{\Lambda}=diag\left[\varepsilon_{11},\varepsilon_{11},\varepsilon_{11}\right],\ \boldsymbol{\Lambda}_1=diag\left[\varepsilon'_{11},\varepsilon'_{11},\varepsilon'_{11}\right] \tag{112}$$

$$\boldsymbol{\Pi}=diag\left[\mu_{11},\mu_{11},\mu_{11}\right],\ \boldsymbol{\Pi}_1=diag\left[\mu'_{11},\mu'_{11},\mu'_{11}\right] \tag{113}$$

$$\boldsymbol{\Phi}=\boldsymbol{\Psi}=\begin{bmatrix}1 & 0 & 0\\ 0 & 1 & 0\\ 0 & 0 & 1\end{bmatrix} \tag{114}$$

We can see from the above equations that there is only one eigen-space in isotropic one-order dispersive medium, which is a triple-degenerate one, and the space structure is the following

$$\boldsymbol{W}_{mag}=W_1^{(3)}\left[\boldsymbol{\phi}_1,\boldsymbol{\phi}_2,\boldsymbol{\phi}_3\right],\ \boldsymbol{W}_{ele}=W_1^{(3)}\left[\boldsymbol{\varphi}_1,\boldsymbol{\varphi}_2,\boldsymbol{\varphi}_3\right] \tag{115}$$

where, $\boldsymbol{\phi}_1^*=\frac{\sqrt{3}}{3}\{1,1,1\}^T,\ \boldsymbol{\varphi}_1^*=\frac{\sqrt{3}}{3}\{1,1,1\}^T,\ \xi_1=1$. Thus the eigen-qualities and eigen-operators of isotropic one-order dispersive medium are known as same as Eqs. (88) – (90). The equations of electromagnetic wave in one-order dispersive medium become

$$\left(\partial_x^2+\partial_y^2+\partial_z^2\right)E_1^*=\frac{1}{c^2}\nabla_{tt}E_1^*+\left(\mu_{11}\varepsilon_{11}^{(1)}+\mu_{11}^{(1)}\varepsilon_{11}\right)\nabla_{ttt}E_1^* \tag{116}$$

$$\left(\partial_x^2+\partial_y^2+\partial_z^2\right)H_1^* = \frac{1}{c^2}\nabla_{tt}H_1^* + \left(\mu_{11}\varepsilon_{11}^{(1)} + \mu_{11}^{(1)}\varepsilon_{11}\right)\nabla_{ttt}H_1^* \tag{117}$$

in which $c$is the velocity of electromagnetic wave

$$c = \frac{1}{\sqrt{\mu_{11}\varepsilon_{11}}} \tag{118}$$

Now, we discuss the propagation laws of a plane electromagnetic wave in x-axis. In this time, Eq.(116) becomes

$$\frac{\partial}{\partial x^2}E_1 = \frac{1}{c^2}\nabla_{tt}E_1 + \left(\mu_{11}\varepsilon_{11}^{(1)} + \mu_{11}^{(1)}\varepsilon_{11}\right)\nabla_{ttt}E_1 \tag{119}$$

Let

$$E_1 = A\exp\left[i(kx-\omega t)\right] \tag{120}$$

Substituting the above into Eq.(119), we have

$$k^2 = \frac{\omega^2}{c^2} - i\omega^3\left(\mu_{11}\varepsilon_{11}^{(1)} + \mu_{11}^{(1)}\varepsilon_{11}\right) \tag{121}$$

From the above, we can get

$$\bar{k} = k_1 + ik_2 \tag{122}$$

Where

$$k_1 = \frac{\omega}{c}\left[\frac{1+\left(1+c^4\left(\mu_{11}\varepsilon_{11}^{(1)}+\mu_{11}^{(1)}\varepsilon_{11}\right)^2\omega^2\right)^{\frac{1}{2}}}{2}\right]^{\frac{1}{2}},\quad k_2 = \frac{\omega}{c}\left[\frac{-1+\left(1+c^4\left(\mu_{11}\varepsilon_{11}^{(1)}+\mu_{11}^{(1)}\varepsilon_{11}\right)^2\omega^2\right)^{\frac{1}{2}}}{2}\right]^{\frac{1}{2}}.$$

Then, the solutions of electromagnetic waves are

$$E_1 = Ae^{-k_2x}\cdot e^{i(k_1x-\omega t)} = \bar{A}\cdot e^{i(k_1x-\omega t)} \tag{123}$$

It is an attenuated sub-waves.

8. ELECTROMAGNETIC WAVES IN CHIRAL MEDIA

8.1. The constitutive equation of chiral media

The constitutive equations of chiral media are the following

$$D = \varepsilon \cdot E - \chi \cdot \nabla_t H \quad (124)$$

$$B = \chi \cdot \nabla_t E + \mu \cdot H \quad (125)$$

where χ is the matrix of chirality parameter, and a symmtric one.

Substituting Eqs. (5) and (6) into Eqs. (124) and (125), respectively, and multiplying them with the transpose of modal matrix in the left, we have

$$D^* = \Phi^T \varepsilon \Phi E^* - \Phi^T \chi \Psi \nabla_t H^* \quad (126)$$

$$B^* = \Psi^T \chi \Phi \nabla_t E^* + \Psi^T \mu \Psi H^* \quad (127)$$

Let $\Gamma = \Psi T \chi \Phi$, that is a coupled chiral matrix, and using Eq. (4), we have

$$D^* = \Lambda E^* - \Gamma^T \nabla_t H^* \quad (128)$$

$$B^* = \Gamma^T \nabla_t E^* + \Pi H^* \quad (129)$$

For most chiral, $\Gamma = diag[\varsigma 1, \varsigma 2, \varsigma 3]$. Then we have

$$D_i^* = \lambda_i E_i^* - \varsigma_i \nabla_t H_i^* \quad (130)$$

$$B_i^* = \varsigma_i \nabla_t E_i^* + \gamma_i H_i^* \quad (131)$$

Eqs.(130) and (131) are just the modal constitutive equations for anisotropic chiral media.

8.2. Eigen equations of electromagnetic waves in chiral media

Substituting Eqs. (130) and (131) into Eqs. (23) and (24), respectively, we have

$$\{\Delta_i^*\}E_i^* = -\nabla_t\{\varphi_i\}(\varsigma_i\nabla_t E_i^* + \gamma_i H_i^*) \quad i=1,2,3 \tag{132}$$

$$\{\Delta_i^*\}H_i^* = \nabla_t\{\phi_i\}(\lambda_i E_i^* - \varsigma_i\nabla_t H_i^*) \quad i=1,2,3 \tag{133}$$

From them, we can get

$$\square_i^* E_i^* + \xi_i\varsigma_i^2\nabla_{tttt}E_i^* + (2\varsigma_i\partial_i^* + \xi_i\lambda_i\gamma_i)\nabla_{tt}E_i^* = 0 \quad i=1,2,3 \tag{134}$$

$$\square_i^* H_i^* + \xi_i\varsigma_i^2\nabla_{tttt}H_i^* + (2\varsigma_i\partial_i^* + \xi_i\lambda_i\gamma_i)\nabla_{tt}H_i^* = 0 \quad i=1,2,3 \tag{135}$$

where $\xi i=\{\phi i\}T\cdot\{\varphi i\}=1$, $\partial i=\{\Delta * i\}T\cdot\{\varphi i\}$. Eqs.(134) and (135) are just equations of electric field and magnetic field for chiral media.

8.3. Applications

In this section, we discuss the propagation laws of electromagnetic waves in an isotropic chiral medium. The material tensors in Eqs.(124) and (125) are represented by the following matrices

$$\boldsymbol{\varepsilon} = \begin{bmatrix} \varepsilon_{11} & 0 & 0 \\ 0 & \varepsilon_{11} & 0 \\ 0 & 0 & \varepsilon_{11} \end{bmatrix}, \ \boldsymbol{\mu} = \begin{bmatrix} \mu_{11} & 0 & 0 \\ 0 & \mu_{11} & 0 \\ 0 & 0 & \mu_{11} \end{bmatrix}, \ \boldsymbol{\chi} - \begin{bmatrix} \chi_{11} & 0 & 0 \\ 0 & \chi_{11} & 0 \\ 0 & 0 & \chi_{11} \end{bmatrix} \tag{136}$$

The eigen values and eigen vectors of those matrix are the following

$$\boldsymbol{\Lambda} = diag[\varepsilon_{11}, \varepsilon_{11}, \varepsilon_{11}], \ \boldsymbol{\Pi} = diag[\mu_{11}, \mu_{11}, \mu_{11}], \ \boldsymbol{\Gamma} = diag[\chi_{11}, \chi_{11}, \chi_{11}] \tag{137}$$

$$\boldsymbol{\Phi} = \boldsymbol{\Psi} = \begin{bmatrix} 1 & 0 & 0 \\ 0 & 1 & 0 \\ 0 & 0 & 1 \end{bmatrix} \tag{140}$$

We can see from the above equations that there is only one eigen-space in isotropic medium, which is a triple-degenerate one, and the space structure is the following

$$\boldsymbol{W}_{mag} = W_1^{(3)}[\boldsymbol{\phi}_1, \boldsymbol{\phi}_2, \boldsymbol{\phi}_3], \ \boldsymbol{W}_{ele} = W_1^{(3)}[\boldsymbol{\varphi}_1, \boldsymbol{\varphi}_2, \boldsymbol{\varphi}_3] \tag{141}$$

where, $\phi_1^* = \frac{\sqrt{3}}{3}\{1,1,1\}^T$, $\varphi_1^* = \frac{\sqrt{3}}{3}\{1,1,1\}^T$, $\xi_1 = 1$. Then the eigen-qualities and eigen-operators of isotropic chiral medium are respectively shown as follows

$$E_1^* = \frac{\sqrt{3}}{3}(E_1 + E_2 + E_3), \quad H_1^* = \frac{\sqrt{3}}{3}(H_1 + H_2 + H_3) \tag{142}$$

$$\Box_1^* = \frac{1}{3}\left[-\left(\partial_x^2 + \partial_y^2 + \partial_z^2\right)\right], \quad \partial_1^* = \frac{\sqrt{3}}{3}\left(\partial_x + \partial_y + \partial_z\right) \tag{143}$$

So, the equations of electromagnetic wave in isotropic chiral medium become

$$\left(\partial_x^2 + \partial_y^2 + \partial_z^2\right)E_1^* = \chi_{11}^2 \nabla_{tttt} E_1^* + \left[2\sqrt{3}\chi_{11}\left(\partial_x + \partial_y + \partial_z\right) + \frac{1}{c^2}\right]\nabla_{tt} E_1^* \tag{144}$$

$$\left(\partial_x^2 + \partial_y^2 + \partial_z^2\right)H_1^* = \chi_{11}^2 \nabla_{tttt} H_1^* + \left[2\sqrt{3}\chi_{11}\left(\partial_x + \partial_y + \partial_z\right) + \frac{1}{c^2}\right]\nabla_{tt} H_1^* \tag{145}$$

where, $c$is the velocity of electromagnetic wave

$$c = \frac{1}{\sqrt{\mu_{11}\varepsilon_{11}}} \tag{146}$$

Now, we discuss the propagation laws of a plane electromagnetic wave in x-axis. In this time, Eq.(142) becomes

$$\frac{\partial}{\partial x^2} E_1 = \chi_{11}^2 \nabla_{tttt} E_1 + \left[2\sqrt{3}\chi_{11}\frac{\partial}{\partial x} + \frac{1}{c^2}\right]\nabla_{tt} E_1 \tag{147}$$

Let

$$E_1 = A\exp\left[i(kx - \omega t)\right] \tag{148}$$

Substituting the above into Eq.(145), we have

$$k^2 = \left(\frac{1}{c^2} + i2\sqrt{3}\chi_{11}k\right)\omega^2 - \chi_{11}^2\omega^4 \tag{149}$$

or

$$k^2 - i2\sqrt{3}\chi_{11}\omega^2 k + \left(\chi_{11}^2\omega^2 - \mu_{11}\varepsilon_{11}\right)\omega^2 = 0 \tag{150}$$

when $\omega^2 - \dfrac{\mu_{11}\varepsilon_{11}}{\chi_{11}^2} < 0$

By Eq.(148), we have

$$k_1 = k_1' + ik_1'' = i\chi_{11}\omega^2\left[\sqrt{3} + (x+iy)\right] \tag{149}$$

$$k_2 = k_2' + ik_2'' = i\chi_{11}\omega^2\left[\sqrt{3} - (x+iy)\right] \tag{150}$$

From them, we can get

$$k_1' = -\chi_{11}\omega^2 y\ ,\ k_1'' = \left(\sqrt{3}+x\right)\chi_{11}\omega^2 \tag{151}$$

$$k_2' = \chi_{11}\omega^2 y\ ,\ k_2'' = \left(\sqrt{3}-x\right)\chi_{11}\omega^2 \tag{152}$$

where, $x = \dfrac{\sqrt{2}}{2}\sqrt{3 + \left[9 + \dfrac{\left(\dfrac{\mu_{11}\varepsilon_{11}}{\chi_{11}^2} - \omega^2\right)^2}{\omega^4}\right]^{\frac{1}{2}}}\ ,\ y = \dfrac{\sqrt{2}}{2}\sqrt{-3 + \left[9 + \dfrac{\left(\dfrac{\mu_{11}\varepsilon_{11}}{\chi_{11}^2} - \omega^2\right)^2}{\omega^4}\right]^{\frac{1}{2}}}$

Then, the solution of electromagnetic waves is the following

$$E_1 = A_1 e^{-k_1'' x} \cdot e^{i(k_1' x - \omega t)} + A_2 e^{-k_2'' x} \cdot e^{i(k_2' x - \omega t)} \tag{153}$$

It is composed of two attenuated sub-waves.

2. when $\omega^2 - \dfrac{\mu_{11}\varepsilon_{11}}{\chi_{11}^2} > 0$

By Eq.(148), we have

$$k_1 = k_1' + ik_1'' = i\chi_{11}\omega^2\left[\sqrt{3} + \sqrt{3 + \frac{1}{\omega^2}\left(\omega^2 - \frac{\mu_{11}\varepsilon_{11}}{\chi_{11}^2}\right)}\right] \tag{154}$$

$$k_2 = k_2' + ik_2'' = i\chi_{11}\omega^2\left[\sqrt{3} - \sqrt{3 + \frac{1}{\omega^2}\left(\omega^2 - \frac{\mu_{11}\varepsilon_{11}}{\chi_{11}^2}\right)}\right] \tag{155}$$

$$k_1' = 0\,,\; k_1'' = \chi_{11}\omega^2\left[\sqrt{3} + \sqrt{3 + \frac{1}{\omega^2}\left(\omega^2 - \frac{\mu_{11}\varepsilon_{11}}{\chi_{11}^2}\right)}\right] \tag{156}$$

$$k_2' = -\chi_{11}\omega^2\left[\sqrt{3 + \frac{1}{\omega^2}\left(\omega^2 - \frac{\mu_{11}\varepsilon_{11}}{\chi_{11}^2}\right)} - \sqrt{3}\right],\; k_2'' = 0 \tag{157}$$

Then, the solution of electromagnetic waves is the following

$$E_1 = A_1 e^{-k_1'' x} e^{-i\omega t} + A_2 e^{i(k_2' x - \omega t)} \tag{158}$$

It is seen that there only exists an electromagnetic sub-wave in opposite direction.

3. when $\omega^2 - \frac{\mu_{11}\varepsilon_{11}}{\chi_{11}^2} = 0$

By Eq.(148), we have

$$k = k' + ik'' = i2\sqrt{3}\chi_{11}\omega^2 \tag{159}$$

where

$$k' = 0\,,\; k'' = 2\sqrt{3}\chi_{11}\omega^2 \tag{160}$$

Then, the solution of electromagnetic waves is the following

$$E = Ae^{-k'' x} \cdot e^{-i\omega t} \tag{161}$$

No electromagnetic sub-waves exist now.

9. CONCLUSION

In this chapter, we construct the standard spaces under the physical presentation by solving the eigen-value problem of the matrixes of dielectric permittivity and magnetic permeability, in which we get the eigen dielectric permittivity and eigen magnetic permeability, and the corresponding eigen vectors. The former are coordinate-independent and the latter are coordinate-dependent. Because the eigen vectors show the principal directions of electromagnetic media, they can be used as the standard spaces. Based on the spaces, we get the modal equations of electromagnetic waves for anisotropic media, bi-anisotropic media, dispersive medium and chiral medium, respectively, by converting the classical Maxwell's vector equation to the eigen Maxwell's scalar equation, each of which shows the existence of an electromagnetic sub-wave, and its propagation velocity, propagation direction, polarization direction and space pattern are completely determined in the equations. Several novel results are obtained for anisotropic media. For example, there is only one kind of electromagnetic wave in isotropic crystal, which is identical with the classical result; there are two kinds of electromagnetic waves in uniaxial crystal; three kinds of electromagnetic waves in biaxial crystal and three kinds of distorted electromagnetic waves in monoclinic crystal. Also for bi-anisotropic media, there exist two electromagnetic waves in Dzyaloshinskii's bi-anisotropic media, and the electromagnetic waves in bi-anisotropic medium will go faster duo to the bi-coupling between electric field and magnetic one. For isotropic dispersive medium, the electromagnetic wave is an attenuated sub-waves. And for chiral medium, there exist different propagating states of electromagnetivc waves in different frequency band, for example, in low frequency band, the electromagnetic waves are composed of two attenuated sub-waves, in high frequency band, there only exists an electromagnetic sub-wave in opposite direction, and in the critical point, no electromagnetic can propagate. All of these new theoretical results need to be proved by experiments in the future.

REFERENCES

1. A. Agyei, J. Birman, 1990 On the linear magnetoelectric effect. Journal of Physics: Condensed Matter, 2 13 3007 3020 , 0953-8984
2. D. Astrov, 1960 The magnetoelectric effect in antiferromagnetics. Soviet Physics, Journal of Experimental and Theoretical Physics, 38 11 984 985 , 0038-5646
3. C. Athanasiadis, S. Giotopoulos, 2003 The Atkinson-Wilcox expansion theorem for electromagnetic chiral waves. Applied Mathematics Letters, 16 5 675 681 , 0893-9659
4. C. Athanasiadis, G. Roach, I. Stratis, 2003 A time domain analysis of wave motions in chiral materials. Mathematische Nachrichten, 250 1 3 16 , 1522-2616
5. Y. Benveniste, 1995 Magnetoelectric effect in fibrous composites with piezoelectric and piezomagnetic phases. Physical Review B 51 9 424 427 , 1098-0121
6. L. Bracke, M. Van Vliet, 1981 A broadband magneto-electric transducer using a composite material. International Journal of Electronics, 51 2 255 262 , 0020-7217
7. M. Bui, S. Stuchly, G. Costache, 1991 Propagation of transients in dispersive dielectric media. IEEE Transactions on Microwave Theory and Techniques, 39 7 1165 1171 , 0018-9480
8. G. Cohen, 2002 Higher-order numerical methods for transient wave equations, Springer, 354041598 Berlin
9. I. Dzyaloshinskii, 1959 On the magneto-electrical effect in antiferromagnets. Soviet Physics, Journal of Experimental and Theoretical Physics, 10 3 628 669 , 0038-5646
10. S. Dvorak, D. Dudley, 1995 Propagation of ultra-wide-band electromagnetic pulses through dispersive media. IEEE Transactions on Electromagnetic Compatibility, 37 1 192 200 , 0018-9375

11. O. Gandhi, C. Furse, 1997 Currents induced in the human body for exposure to ultrawideband electromagnetic pulses. IEEE Transactions on Electromagnetic Compatibility, 39 2 174 180 , 0018-9375

12. S. Guo, 2009 An eigen theory of electromagnetic waves based on the standard spaces. International Journal of Engineering Science, 47 3 405 412 , 0020-7225

13. S. Guo, 2009 An Eigen theory of Static Electromagnetic Field for Anisotropic Media. Applied Mathematics and Mechanics, 30 5 643 652 , 0253-4827

14. S. Guo, 2010 A second-order eigen theory for static electromagnetic fields. Journalof Modern Physics, 1 2 100 107 , 2153-1196

15. S. Guo, 2010 An eigen theory of electro-magnetic acoustic waves in magnetoelectroelastic media. Acta Mechanica, 211 2 173 180 , 0001-5970

16. S. Guo, 2010 An fully dynamical theory of piezoelectromagnetic waves. Acta Mech., 215 3 335 344 , 0001-5970

17. Z. Haba, 2004 Green functions and propagation of waves in strongly inhomogeneous media. Journal of Physis A: Mathematical and General, 37 9 9295 9302 , 0305-4470

18. G. Harshe, J. Dougherty, R. Newnham, 1993 Theoretical modelling of multilayer magnetoelectric composites. International Journal of Applied Electromagnetic Matter, 4 2 145 159 , 1383-5416

19. D. Jaggard, A. Mickelson, C. Papas, 1979 On electromagnetic waves in chiral media. Applied Physics A: Materials Science and Processing, 18 2 211 216 , 0947-8396

20. J. Jin, M. Zunoubi, K. Donepudi, W. Chew, 1999 Frequency-domain and time-domain finite element solution of Maxwell's equations using spectral Lanczos decomposition method.

Computer Methods in Applied Mechanics and Engineering, 169 2 279 296 , 0045-7825

21. J. Kong, 1986 Electromagnetic wave theory, John Wiley & Sons, 978-0-96681-439-2 New York

22. A. Lakhtakia, 1994 Beltrami fields in chiral media. World Scientific, 9-81021-403-0

23. A. Lakhtakia, V. K. Varadan, V. V. Varadan, 1989 Time-harmonic electromagnetic fields in chiral media. Springer-Verlag, 3-54051-317-5

24. L. Landau, I. Lifshitz, 1960 Electrodynamics of Continuous Media. Pergamon Press, 0-44486-165-3

25. R. Lee, N. Madsen, 1990 A mixed finite element formulation for Maxwell's equations in the time domain. Journal Computational Physics, 88 2 284 304 , 0021-9991

26. I. Lindell, A. Sihvola, S. Tretyakov, A. Viitanen, 1994 Electromagnetic Waves in Chiral and Bi-isotropic Media. Artech House, ISBN, Boston

27. I. Lindell, A. Sihvola, S. Tretyakov, A. Viitanen, 1994 Electromagnetic Waves in Chiral and Bi-isotropic Media., Artech House, 1-59829-320-6

28. J. Li, Y. Chen, 2006 Analysis of a time-domain finite element method for 3D Maxwell's equations in dispersive media. Computer Methods in Applied Mechanics and Engineering, 195 11 4220 4229 , 0045-7825

29. Q. Liu, G. Fan, 1999 Simulation of GPR in dispersive media using a frequency-dependent PSTD algorithm. IEEE Transactions on Geoscience and Remote Sensing, 37 5 2317 2324 , 0196-2892

30. T. Lu, P. Zhang, W. Cai, 2004 Discontinuous Galerkin methods for dispersive and lossy Maxwell's equations and PML boundaryconditions. Journal Computational Physics, 200 4 549 580 , 0021-9991

31. P. Monk, 1992 A comparison of three mixed methods for time-dependent Maxwell's equations. SIAM Journal of Science Statistics Computer, 13 10 1097 1122 , 0960-3174

32. P. Monk, 2003 Finite Element methods for Maxwell's equations, Clarendon Press, 0-19850-888-3

33. T. Pradhan, 1993 Magnetic charge induced by electric charge in a magneto-electric medium. Europhysics Letters, 21 7 971 973 , 0295-5075

34. A. Van Run, D. Terrell, J. Scholing, 1974 An in situ grown eutectic magnetoelectric composite material. Journal of Materials Science, 9 11 1710 1714 , 0022-2461

35. V. Yakhno, T. Yakhno, M. Kasap, 2006 A novel approach for modeling and simulation of electromagnetic waves in anisotropic dielectrics. Internatioal Journal of Solids and Structures, 43 12 6261 6276 , 0020-7683

Chapter 4

Results and Prospects of Experimental Validation of the Use of Electromagnetic Waves at Terahertz Frequencies of Cellular Metabolites in Clinical Practice

V. F. Kirichuk

Department of Normal Physiology, Saratov State Medical University, Saratov, Russia

ABSTRACT

The paper presents the results of experimental studies of the effect of irradiation with electromagnetic waves terahertz frequencies of molecular spectrum of emission and absorption of nitric oxide and atmospheric oxygen on animals in a state of acute and chronic immobilization stress, a condition regarded as a model of violations in different parts of the vascular system including microcirculation specific for different parts of the vascular system, including the microcirculation characteristic of various diseases of the cardiovascular system.

Keywords: Terahertz Waves, Microcirculation, Acute Stress

1. INTRODUCTION

Electromagnetic radiation in the terahertz frequency range is now increasingly beginning to be applied mainly as a correction factor of microcircular violations analysis of the literature, and clinical observations suggest that MM-wave is effective for many chronic diseases as monotherapy in combination with other physical methods and medication, strengthening and consolidating the therapeutic effect [1] -[6] .

Terahertz frequency range (TGCH) lies on the border between electronics and photonics from 100 GGts to 10 TGts (1 THz = 10 Z GHz) or in wavelengths from 3 mm to 30 mm . It is found that the considered wavelength range is used by living organisms for communication and control, and that the organisms themselves emit vibrations millimeter waves excited by the body when exposed to TGCH exposure, and to some extent mimic the signals of internal communication and control (data communication) of biological objects. The result is reduced by the normal range and power of the radiation characteristic of the organism, thus representing that a range of frequencies do not qualitatively change the body, but can be adjusted to normalize its physiological state within specific to the biological species.

It is believed that the reactivity of molecules excited terahertz quantum will also include what TGCH radiation easily penetrates through clothing and skin to the muscle man.

On the other hand, the terahertz frequency range increasingly attracts attention, because of this frequency range centered emission and absorption spectra critical cellular metabolites—NO_2, CO_2, CO, OH, etc. [2] .

Naturally, the greatest interest is the molecular spectrum of electromagnetic waves of radiation and absorption of nitric oxide. It is one of the most important biological mediators involved in many physiological and pathophysiological processes, acting as a second messenger in most cells. In particular, nitric oxide is involved in the implementation of many important physiological functions, such as vasodilation, bronchodilation, neurotransmission, platelet aggregation, immune system reactions, the regulation of smooth muscle tone, and others [6] .

Research and development of methods to control the secretion of maintaining the level of physiological reactivity and concentration of endogenous nitric oxide in the cells, organs and the whole body are scientific and of practical interest for theoretical and clinical medicine. In this regard, currently, intensive search for methods is constantly conducted to create pharmaceutical preparations regulating the secretion and physiological activity of nitric oxide molecules in cells, organs, and body as a whole. However, pharmacological regulation altered physiological functions in the body may be accompanied by the appearance of some persistent undesirable side effects, and in some cases by severe allergic reactions, which limit the breadth of application of medicaments.

This dictates the need to find non-invasive physical controllers formation and secretion of endogenous nitric oxide on the basis of natural physiological processes. The use of low-intensity terahertz radiation at frequencies of molecular spectrum of emission and absorption of nitric oxide and atmospheric oxygen is promising in this respect.

According to the contract No. 005/037/002 on September 25, 2001, Saratov State University and the Ministry of Health and Social Development of the Russian Federation at the Department of Physiology named by I. A. Chuevsky within the industry research program No. 9 “etiology and pathogenesis, diagnosis and treatment of diseases of the blood” are working on “studies on the effect of complex biological systems of electromagnetic waves at frequencies of molecular emission and absorption spectra of substances involved in metabolic processes”. Agreements were signed on the joint scientific and technical activities between Saratov State Medical University and JSC “Central Research Institute of Measuring Equipment” (Saratov)—contract No. 67 on January 31, 2000, contract No. 73 on April 16, 2002 , contract No. 14 on March 28, 2006, as well as a research center for biophotonics

Institute of Biomedical Engineering and Technology health and the Institute of Advanced technology (Shenzhen), Chinese Academy of Sciences—the agreement on March 2, 2010. In addition, studies were carried out under the program of Medical Sciences' scientific medical research Volga region for 2008-2010 on study of the behavioral responses, and the nature of changes in blood flow in the great vessels of blood rheology, microcircular and coagulation mechanisms of homeostasis in biological objects in a state of acute and chronic immobilization stress radiopulse under the influence of radiation of high power and different terahertz frequencies (135 - 250 GHz).

2. MATERIAL AND METHODS

The studies were conducted on 2175 adult white rats of both sexes weighing 150 - 200 grams in conditions of acute and chronic stress. Used advanced studies of vascular and endovascular mechanisms of microcirculation. Exposure of rats was terahertz electromagnetic waves at frequencies of molecular emission spectrum absorbtion nitric oxide (150.176 - 150.664 GHz) and atmospheric oxygen (129 GHz) Apparatus of "Orbit".

3. RESULTS AND DISCUSSION

Based on studies conducted by staff of the Department of Physiology of the normal, I. A. Chuevsky experimentally uses clinical practice, the electromagnetic waves at different frequencies of molecular spectrum of emission and absorption of nitric oxide and oxygen in the intravascular and vascular units microcirculation, the functional activity of the endocrine glands, states lipid peroxidation and antioxidant activity of blood, the behavioral responses of animals, and antistress activity of the body.

Investigations were carried out in two directions, the in vitro and in vivo in two models of breach of these indicators' homeostatic responses of animals (white rats of both sexes), and acute and long-term stress, which were the typical pathological process in various diseases of the cardio—vascular system, malignant diseases of various inflammatory processes.

In vitro studies on platelet-rich plasma in patients with unstable angina showed a normalizing effect of electromagnetic waves of extremely high and terahertz frequency band of the molecular spectrum of emission and absorption of nitric oxide on platelet hemostasis, whose intravascular component of the microcirculation in the comparative evaluation of different modes of electromagnetic millimeter waves for the first time identified that the view EHF radiation was the most efficient for the normalization of activity and platelet aggregation in an in vitro (amplitude nomodulirovannoe effect on the frequency of 42.2 GHz). First, it studied the effects of platelet-THF-radiation at frequencies of molecular spectrum of emission and absorption of nitric oxide 150.176 - 150.664 GHz mode. TGCH effects on specified frequencies of molecular spectrum of emission and absorption of nitric oxide caused the most pronounced normalization of functional activity of platelets in patients with unstable angina (15-minute exposure, and continuous signal generation). A comparative analysis of platelet effects of EHF-exposure and impacts on the frequency spectrum of the molecular absorption and emission of nitrogen oxide shows a more pronounced effect of the latter on the functional activity (activation and aggregation) of platelets in patients with unstable angina. For the first time, it shows the possibility of information impact of intact platelets of healthy subjects of patients with unstable angina platelets subjected TCHG—radiation at frequencies of molecular spectrum of emission and absorption of nitric oxide 150.176 - 150.664 GHz.

The effect of electromagnetic radiation in the frequency range terahertz molecular spectrum of emission and absorption of nitric oxide 240 GHz on platelet aggregation, blood rheology, coagulation potential and fibrinolytic activity of the blood in patients with unstable angina in the conditions shows that irradiation of terahertz waves at a frequency of molecular spectrum of emission and absorption nitric oxide 240 GHz in the in vitro leads to the normalization of indicators characterizing platelet aggregation, blood rheology, coagulation and fibrinolytic potential of the activity of the blood in patients with unstable angina.

The dependence of the efficiency of the exposure effects of electromagnetic radiation in the terahertz frequency range of molecular absorption and emission of nitrogen oxide 240 GHz under in vitro conditions for the changed parameters of platelet function, the rheological properties of blood (blood viscosity, aggregation and deformability of red blood cells), and procoagulant, anticoagulant and fibrinolytic activity of the blood in the natural electromagnetic field of patients with unstable angina most effective in restoring altered blood parameters is a 15-minute exposure mode.

There is a dependence of the efficiency of the exposure effects of electromagnetic radiation in the terahertz frequency range of molecular nitrogen oxide in the in vitro conditions for the changed parameters of platelet function, rheology coagulation potential and fibrinolytic activity of the blood found in crossed magnetic fields and electrolytic, patients with unstable angina. The most effective in restoring altered blood parameters are 15- and 30-minute exposure modes.

For correction of impaired platelet function ballroom unstable angina in vitro used mode terahertz radiation waves at a frequency of 400 GHz nitric oxide. The resulting positive result indicates the need for further study of the use of this mode of exposure to the changed parameters of the microcirculation in patients with cardiology. The rats are able to oxidative stress simulating similar changes in endothelial dysfunction in patients with unstable angina were increased platelet aggregation, blood rheology change—increased viscosity, erythrocyte aggregation and deformability blocked, a marked change in the hypercoagulable hemostasis and depression fibrinolytic activity of blood.

Found that the effect of the normalizing effect of THz radiation on the frequency spectrum of the molecular absorption and emission of nitrogen oxide 240 GHz on the changed parameters of platelet aggregation, rheological properties of the coagulation and fibrinolytic potential of blood is based on the activation of endothelial NO synthase, responsible for the physiological synthesis of nitric oxide.

Thus, the experiment proved the prospect of using terahertz radiation at a frequency range of molecular absorption and emission of nitrogen oxide 240 GHz in clinical practice for the normalization of the changed parameters of platelet aggregation, rheological properties of the coagulation and fibrinolytic potential of blood in patients with cardiology. There is a possibility of further study of terahertz radiation at a frequency range of molecular absorption and emission of nitrogen oxide 400 GHz to justify its use in clinical practice. The influence of electromagnetic radiation TGCH -band frequency spectrum of the molecular emission and absorption of atmospheric oxygen 129 GHz on the rheological properties of blood patchy, with 15- and 30-minute exposure is set lower the viscosity of whole blood, and at 60-minute exposure of its increase in the groups of patients with stable angina and almost healthy individuals. In this case, two modes of exposure cause opposite changes aggregation activity of the deformability of red blood cells and the ability of their membranes, and at 60-minute exposure—increasing the ability of red blood cell aggregation and a slight increase in ductility of red blood cell membranes.

The influence of electromagnetic radiation terahertz frequency MSIP nitric oxide 150.176 - 150.664 GHz on the rheological properties of blood varies depending on exposure and radiation exposure resulted in an increase, as a rule, a more pronounced increase in the viscosity of whole blood of patients with stable angina and healthy individuals. A similar pattern was revealed with an increase in the radiation power. For all modes, exposure frequency EMR MSIP nitric oxide changes in functional parameters of red blood cells were found in different directions increase the ability of red blood cell aggregation and deformability decrease in membrane.

The influence of electromagnetic radiation frequencies 42.19 GHz and 53 GHz to 54 rheological parameters in patients with stable angina and healthy individuals is to increase the viscosity of whole blood by increasing the ability of the aggregation of red blood cells in patients with stable angina and in healthy individuals.

There is a significant dependence of the effect of EMI MMD blood cell peripheral blood at baseline Maximum severity of the rheological parameters of whole blood irradiation EMI MMD is determined by an optimal ratio of blood red cells, white cells and platelets.

Comparative analysis of the impact of EMR on the MMD "classical" frequencies (42.19 - 53.54 GHz) frequencies and MSIP nitric oxide and atmospheric oxygen showed that the severity of the effect of these frequencies EHF EMR on the rheological properties of blood are increased in the direction of 42, 19, 129, 53, 54, 150.176 - 150.664 GHz.

The severity of changes in the shape size, aggregating capacity of red blood cells, the plasticity of their membranes and cytometric changes due to the growth of the power supply and an increase in the exposure of EMR MMD all studied frequencies and for all modes of exposure. Effect of EMP MMD different frequencies on the viscosity of whole blood depends on the extent of atherosclerosis in the coronary arteries and the presence of hypertension increase with the number of coronary arteries, involved in the atherosclerotic process, increase efficiency index for the viscosity of whole blood erythrocyte aggregation and deformability of the membranes, the presence of hypertension patient leads to significantly higher efficiency index for the viscosity of whole blood, red blood cell aggregation and deformability of the ability of their membranes.

Identified "Information Impact" TGCH—irradiated whole blood in patients with stable angina and healthy individuals. Red blood cells are involved in the process of implementing one introduced EMI MMD "information signal" during irradiation frequency EMR 42, 19, 53, 54 and 150.176 - 150.664 GHz.

The effectiveness of the influence of electromagnetic radiation THF-MSIP frequency range 240 GHz of nitric oxide in the in vitro disturbed viscous properties of whole blood aggregation ability and deformability of red blood cells in patients with unstable angina who are in the natural electromagnetic field depends on the duration of exposure. 15-minute exposure mode is most effective in restoring the disturbed viscosity properties of blood and red blood cell aggregation and deformability blocked—in 15- and 30-minute exposure.

Normalizing effect of terahertz radiation at a frequency of MSIP nitric oxide 240 GHz in vitro on Ip broken viscous properties of whole blood erythrocyte aggregation and deformability in patients with unstable angina who are in crossed electric and magnet fields, depending on the time of exposure: the restoration of increased viscosity of whole blood is expressed in 15-minute exposure and impaired red blood cell deformability and aggregation—at 15- and 30-minute exposure. The specified mode of exposure is the most effective when compared with the influence of EMR THF 240 GHz at the blood that is in the natural electromagnetic field, as it causes the improvement of the rheological properties of the majority of patients as well.

Normalizing effect on the frequency EMR THF MSIP nitric oxide 240 GHz under in vitro conditions on the broken properties of whole blood viscosity, erythrocyte aggregation and deformability in patients with unstable angina is most marked and maintained for 30 minutes after irradiation. Irradiation at the frequency of MSIP 240 GHz nitric oxide nitric oxide donator Isoketum leads to a more severe degree of recovery altered blood viscosity and deformability of red blood cells in patients with unstable angina compared with non-irradiated Isoketum.

First studied the effects of electromagnetic radiation at frequencies terahertz range of molecular spectrum of nitric oxide 150, 176 - 150, and 664 GHz on the functional activity of platelets in conditions in vivo.

The dependence of the efficiency effects of electromagnetic radiation in the terahertz frequency range of molecular nitrogen oxide 150, 176 - 150, and 664 GHz impaired functional activity of platelets in white male rats in a state of acute immobilization stress on the exposure mode and exposure. Thus, with continuous irradiation of animals in a state of acute immobilization stress, the greatest efficacy of the 30-minute exposure fractional THF-exposure of animals in a state of acute immobilization stress more effectively restores the basic platelet aggregation: 5- and 15-minute exposure more effective than the corresponding duration of continuous irradiation.

First studied the preventive effects of electromagnetic radiation at terahertz frequency range of molecular nitrogen oxide 150, 176 - 150, and 664 GHz stressful violation of platelet function in male rats.

Found that prior immobilization TGCH exposure can prevent the development of stress disorders platelet function. It was found that 5- and 15-minute-exposure modes TGCH to immobilize have a more pronounced effect on platelet aggregation stressful violations compared with the same time of exposure modes on the background of established acute stress. The effectiveness of the 30-minute regime is the same for both methods of application of radiation, and that in both cases there is a complete recovery performance aggregation ability of platelets. First, to show that acute and chronic immobilization stress in male rats and female rats that are in one of the two phases of the estrous cycle or Dioestrus oestus observed violations of the coagulation and fibrinolytic properties of blood. White female rats respond to acute immobilization stress as a function of the estrous cycle in the phase Dioestrus, they are less sensitive to stress factors than in the phase Oestus.

The impact TGCH irradiation at frequencies MSIP nitric oxide 150, 176 - 150, and 664 GHz continuously for 30 minutes in intact animals does not cause changes in the coagulant and fibrinolytic activity of the blood of male rats.

TGCH-radiation at frequencies MSIP nitric oxide 176 - 150.664, and 150 GHz partially or completely normalizes disturbances in coagulation and fibrinolytic activity of blood uderstressed white male rats, depending on the time and mode of exposure.

For the first time found that when animals are irradiated continuously and with acute stress immobilization is most effective mode of 30-minute exposure. Prior to the action of irradiation of a stressor in the continuous mode prevents the development of irregularities in the different phases of blood coagulation and fibrinolysis, with 15-minute irradiation. The fractional mode of exposure of animals with acute immobilization stress is a full restoration of the studied parameters of coagulation hemostasis and fibrinolysis during the 15-minute TGC-influence at frequencies MSIP NO 150.176 - 150.664 GHz. Daily pre-exposure TGCH continuously for 30 minutes of male rats in a state of chronic immobilization stress, completely prevents disturbances in the blood coagulation system in fibrinolysis.

Comparative analysis of the effect of continuous and fractional modes of terahertz radiation at frequencies MSIP nitric oxide 176 - 150.664, and 150 GHz performance coagulative hemostasis and fibrinolysis white male rats in a state of acute immobilization stress, showed that at the 15-minute TGCH-effects in the fractional mode showed complete normalization of disturbed blood coagulation parameters and blood fibrinolytic activity. Continuous exposure mode in the same time repartition showed only partial recovery of indicators of coagulation hemostasis and fibrinolysis.

Effect of pre-and post terahertz radiation at frequencies MSIP nitric oxide 150.176 - 150.664 GHz performance coagulative and fibrinolytic activity of the blood of white male rats in a state of acute immobilization stress is ambiguous. The first shows that the pre-continuous irradiation uderstressed animals within 15 minutes completely prevents the development of disturbances in coagulation hemostasis and fibrinolysis. Continuous exposure mode on the background of established immobilization stress in the same time limit has only a partial recovery of blood coagulation activity indicators and fibrinolytic potential. First discovered features of the effect EMR THF - band frequencies on MSIP NO 150.176 - 150.664 GHz on coagulation and hemostasis fibrinolid white female rats that are in one of the two phases of the estrous cycle Dioesrtus and Oestrus in acute immobilization stress as a function of the irradiation rats females continuously TGCH-waves at frequencies MSIP 176 - 150.664 NO 1 GHz for 15 minutes to an action stressor agent completely prevented the development of disorders of blood coagulation potential and performance of fibrinolysis regardless of the phase of the estrous cycle. Fractional TGCH-radiation at frequencies MSIP NO 150.176 - 150.664 GHz in the same exposure by an acute immobilization stress has only a partial restoration of indicators of coagulation hemostasis and fibrinolysis in female rats in phases Dioesrtus and Oestrus cycle.

It is shown that the most effective in the restoration of disturbed coagulation and fibrinolytic properties of blood in female rats in Dioesrtus phases of the estrous cycle and Oestrus is a preventive mode TGCH continuous irradiation at frequencies MSIP nitric oxide 150.176 - 150.664 GHz for 15 minutes.

Pre continuous 15-minute exposure to EMR THF at frequencies MSIP nitric oxide 150.176 - 150.664 GHz limits the development of acute and chronic stress reactions by reducing the release of glucocorticoids, including corticosterone by the adrenal cortex. This fact is evidence of anti-stress effect of EMI frequency range. For the first time experimentally justified the use of terahertz electromagnetic waves at frequencies of molecular spectrum of emission and absorption of nitric oxide and atmospheric oxygen to restore the disturbed rheological properties of blood and platelet function. The effect of different modes of electromagnetic radiation at terahertz frequencies MSIP nitric oxide 150.176 - 150.664 GHz and atmospheric oxygen at 129.0 GHz aggregation activity of platelets, blood rheology in animals in a state of immobilization stress. It is shown that the effects of terahertz waves at frequencies MSIP nitric oxide 150.176 - 150.664 GHz leads to the normalization of disturbed parameters describing platelet aggregation, blood rheology in white rats. The dependence of the efficiency effects of electromagnetic radiation on the specified range of frequencies 150.176 - 150.664 GHz white rats from the time of exposure to impaired performance of the functional activity of platelets, blood rheology of both males and females with regard to their estrous cycle. The most effective in restoring and preventing the changed parameters of blood rheology in acute stress are continuous and preventive regimens with 15- and 30-minute, and prolonged (chronic) stress 30-minute exposure. Preventive fractional radiation in conditions of acute stress is within 5 minutes limit the development of stress disorders viscosity of blood vessels in the microcirculation.

It is established that the continuous radiation at TGCH impaired by acute stress functional platelet activity of male rats for 30 minutes has a normalizing effect, while the 5- and 15-minute exposure to lead only to a partial recovery of impaired platelet aggregation. Fractional mode TGCH radiation within 5 and 15 minutes restores impaired platelet function caused by acute stress, and more effectively than continuous. Preventive TGCH radiation when exposed for 15 and 30 minutes on male rats is able to prevent the development of characteristic of acute stress reaction violations platelet aggregation. It was revealed that the continuous mode TGCH radiation for 15 and 30 minutes electromagnetic waves at a frequency of molecular spectrum of emission and absorption of atmospheric oxygen 129.0 GHz male rats in a state of acute stress, restores the broken viscous properties of blood, the functional activity of erythrocyte—their aggregation and deformability, as a preventive treatment for the same duration of exposure prevents post stressed violations in terms of blood rheology.

Continuous TGCH radiation within 5 minutes working on white female rats in Dioestrus phases of the estrous cycle and Oestrus in a state of acute stress, electromagnetic waves at a frequency of MSIP atmospheric oxygen 129.0 GHz normalizes the ability of red blood cells in the rat strain in phase Dioestrus 15- and 30-minute exposure fully restore the impaired blood viscosity, and the functional activity of red blood cells regardless of the phase of the estrous cycle. Predictive 5-minute TGCH irradiation of females being Dioestrus barriers development of stress changes the ability of red blood cells to deform, and the 15- and 30-minute exposure completely block post stressed violation of rheological properties of blood regardless of the phase of the estrous cycle.

Continuous exposure to terahertz frequency MSIP atmospheric oxygen 129.0 GHz male rats in acute stress condition for 5 minutes completely restores increased platelet aggregation, 15- and 30-minute modes not only reduced disturbances in platelet function, but also contribute to its further Fractional mode emission reduction by an acute stress sheathe of 15 and 30 minutes is a full recovery of impaired platelet function. Preventive treatment of radiation for 15 and 30 minutes post stressed completely prevents platelet function disorders.

The experimental data provide a basis to extrapolate the use of terahertz radiation at frequencies MSIP nitric oxide 176 - 150.665, 150 GHz and 129.0 GHz atmospheric oxygen in clinical practice to normalize the disturbed parameters of rheological properties of blood and platelet aggregation in patients with different pathologies.

Changes in intravascular component of the microcirculation—qualitative and quantitative composition of red blood cells cause deterioration of the rheological properties of blood during acute and chronic immobilization stress in male rats and female rats that are in one of the two phases of the estrous cycle—Dioestrus or Oestrus. In acute stress immobilization more significantly influence the stress factor for the qualitative and quantitative composition of red blood cells of male rats and female rats under Dioestrus. Female rats in the phase of the estrous cycle of Oestrus were more resistant to the stress factor. Chronic stress violation of quantitative and qualitative composition of red blood cells is similar to those of acute stress.

The absence of the influence of electromagnetic radiation TGCH-band frequencies for MSIP nitric oxide 150.176 - 150.664 GHz on qualitative and quantitative composition of the red blood cells of intact male rats.

TGCH-radiation at frequencies MSIP nitric oxide 150.176 - 150.664 GHz partially or completely normalize post stressed violations in the intravascular component of the microcirculation by restoring the quantitative and qualitative composition of red blood cells, depending on the mode and the exposure time 15-minute exposure mode is the most effective in restoring impaired qualitative and quantitative composition of red blood cells. All the time modes of preventive TGCH-irradiation of animals in a state of acute stress disorders prevent the development of qualitative and quantitative composition of red blood cells. Daily TGCH pre-irradiation for 30 minutes, the animals in a state of chronic stress only restores the mean corpuscular hemoglobin concentration, and the average diameter and erythrocyte aggregation. The features of the influence of electromagnetic radiation TGCH-band frequencies for MSIP nitric oxide 150.176 - 150.664 GHz on qualitative and quantitative composition of red blood cells white female rats that are in one of the two phases of the estrous cycle—Dyuestrus or Oestrus in a state of acute immobilization exposure. Proactive continuous exposure mode for 15 minutes prevents the development of disturbances in the qualitative and quantitative composition of red blood cells during acute stress. Fractional mode of irradiation for 15 minutes prevents the development of qualitative and quantitative composition of red blood cells during acute stress. Fractional mode of irradiation for 15 minutes against the background of established immobilization stress causes the reduction of the qualitative and quantitative composition of red blood cells in female rats only Oestrus in the phase of the estrous cycle. This mode of irradiation is less effective in normalizing stress disturbed blood rheology in female rats in phase Dioestrus.

Comparative analysis of the effect of continuous and fractional modes TGCH irradiation on the frequency of MSIP oxide nitrogen 150.176 - 150.664 GHz on the broken qualitative and quantitative composition of the red blood cells of albino rats of both sexes in a state of acute immobilization stress showed that the fractional modes for 5 and 30 minutes more effective than continuous in those time frames. Preliminary fractional 15-minute exposure mode prevents the development of disturbances in the qualitative and quantitative composition of red blood cells.

Effect of pre- and post-irradiation TGCH at frequencies MSIP nitric oxide 150.176 - 150.664 GHz on qualitative and quantitative composition of the red blood cells of albino rats of both sexes with acute immobilization stress is ambiguous. It is shown that pre- and fractional continuous irradiation for 15 animals and preliminary continuous irradiation for 30 minutes post stressed prevents the development of irregularities in the qualitative and quantitative composition of erythrocytes.

First studied the effects of electromagnetic radiation at terahertz frequencies of molecular spectrum of emission and absorption of nitric oxide 150.176 - 150.664 GHz on morphological and functional microcirculation in various organs of white rats in a state of acute prolonged stress.

It was found that irradiation TGCH capable of restoring the characteristic of acute and long-term immobilization stress microcirculation. This manifests itself in part the normalization of its intravascular component—no signs sladzhsindroma, reducing stasis phenomena: the vascular component—the normalization constant and reducing capillary fragility and extravascular components reducing the severity of edema. It was revealed that the action of the radiation frequency specified in animals in a state of acute immobilization stress observed a significant reduction in ischemic events many internal organs (liver, kidneys, and lungs), and the normalization of blood flow in the brain EMR THF-band frequencies for MSIP nitric oxide 150.176 - 150.664 GHz causes the downward trend of increased concentrations of endothelin- 1 in the blood serum of white male rats in a state of acute immobilization stress. Daily TGCH irradiation for 5 days after each immobilization causes a statistically significant reduction in elevated concentrations of endothelin- 1 in the serum of male rats in a state of prolonged stress. Implementation effect TGCH will in the frequencies of molecular spectrum of emission and absorption of nitric oxide 150.176 - 150.664 GHz on the concentration of endothelin 1 is carried out with the participation of nitric oxide blockade of stress-limiting system of nitric oxide contributes to the development of endothelin 1.

Normalizing effects of terahertz electromagnetic radiation at frequencies of molecular spectrum of emission and absorption of nitric oxide 150.176 - 150.664 GHz on microcirculation in animals in a state of acute and long-term immobilization stress realized by changing the activity of the pituitary adrenal and thyroid glands. TGCH-effects at these frequencies moderately reduces hyperactivity gipofizarnonadpochechnikov and thyroid axis stress response, manifested a lack or decrease in expression characteristic of the acute and long-term stress reaction of morphological changes in the pituitary gland, adrenal glands and the thyroid gland.

First studied the effects of electromagnetic radiation at terahertz frequencies MSIP nitric oxide 150.176 - 150.664 GHz impaired under stress hemodynamics in the great vessels.

Found that the electromagnetic radiation at these frequencies does not affect the linear flow velocity of intact male albino rats.

The dependence of the efficiency effects of electromagnetic radiation in the terahertz frequency range of molecular absorption and emission of nitrogen oxide, 150.176 - 150.664 GHz in the breakdown of the linear velocity of blood flow in white male rats in a state of acute immobilization stress on the mode of exposure. Thus, with continuous exposure of animals in a state of acute immobilization stress 5-, 15-, and 30-min exposure have the same efficiency as increasing time of exposure does not lead to an increase in efficiency, that is, the most effective is a 5-minute regime.

Continuous 30-minute regime TGCH-irradiation of animals in a state of acute immobilization stress more effectively restore the basic hemodynamic parameters compared with the fractional regime TGCH irradiation of the same exposure.

The first time the effectiveness of continuous preventive effect on the frequency range of the electromagnetic spectrum of molecular nitrogen oxide 150.176 - 150.664 GHz stressful violation of linear velocity of blood flow in male rats.

Found that prior immobilization of TGCH irradiation can prevent the development of stress disorders hemodynamics with 5- and 15-minute regimes have equally pronounced effect on the breakdown of the linear velocity of blood flow in the abdominal aorta and femoral artery compared with the same time of continuous exposure modes on the background of the developing acute stress.

The first time the role of endogenous nitric oxide and endothelial NO-synthase involved in the mechanisms of positive coirrecting effects of terahertz radiation at frequencies MSIP nitric oxide 150.176 - 150.664 GHz performance linear velocity of blood flow in the great vessels of white rats in a state of acute immobilization stress, as evidenced by the lack of normalizing effect of terahertz radiation on disturbed hemodynamics when administered blocker endotolialnoy NO synthase L-name white male rats in a state of acute stress.

Under the influence of the course of radiation at frequencies of terahertz waves MSIP nitric oxide 150.176 - 150.664 GHz on a background of prolonged stress in male rats, a partial normalization of blood flow in the major arteries. Reduced blood flow velocity indicates a decrease in vascular tone under the influence of terahertz waves. The decrease in vascular tone is probably due to the increased production of vasoconstrictor substances by endothelial cells. This is confirmed by the results of the study changes in the concentration of stable nitric oxide metabolites—nitrites and endothelin 1—white rats with prolonged immobilization stress.

The data presented indicate that exposure Courses of terahertz waves at a frequency of 150 MSIP nitric oxide 176 - 150.664 GHz capable of restoring the impaired blood flow in the arteries of the white rats with prolonged stress. This makes possible the use of irradiation of these waves to restore hemodynamic disturbances occur in pathological conditions.

Studies show that male rats in a state of acute stress and subjected TGCH irradiation at frequencies of molecular spectrum of emission and absorption of nitric oxide 150.176 - 150.664 GHz, there is a restoration of perfusion index (M), while it is not statistically different from the level of the control group. The animals in this group there is a statistically significant compared to the group of animals in a state of acute stress and not subjected TGCH-irradiation, an increase in FLACSO and the coefficient of variation. It should be noted that there is a tendency to increase the values of FLACSO, and especially the coefficient of variation in the animals in this group compared with the control group, indicating a more intense modulation of the microcirculation and mechanisms of its regulation.

These amplitude-frequency analysis of LDF-grams shows TGCH that after irradiation of male rats in a state of acute immobilization stress is statistically significant increase in the amplitude and endothelial vasomotor fluctuations compared to the group of animals in a state of acute immobilization stress is not affected by TGCH- effects. This is reflected in the increase of endothelial vasodilator activity (activation of basal nitric oxide production) and a decrease in peripheral resistance. Also, there is an increase of pulse amplitude (cardiac) vibrations, indicating an increase in the inflow of arterial blood microcircular channel. All figures are amplitude—frequency analysis of LDF—gram animals in this group did not differ significantly from those of the control group.

Thus, the electromagnetic radiation in the terahertz frequency range of molecular absorption and emission of nitrogen oxide increases 150.176 - 150.664 reduced both basal and stimulated production of nitric oxide by the endothelium in male rats in a state of acute immobilization stress. This is consistent with the results of earlier biochemical studies that showed an increased concentration of nitrite levels in male rats in a state of stress immoblizative influenced electromagnetic waves specified frequency range, indicating the activation of nitroxidergic system. Increasing the role of passive mechanisms of regulation of microcirculation (amplitude increase heart rate, reflecting the increase in the inflow of arterial blood into the bloodstream, and to a lesser extent, the amplitude of the respiratory rhythm) can be regarded as a compensatory response of tissue reperfusion after ischemia. Increasing the role of active and passive modulation mechanisms microcirculation under the influence of electromagnetic radiation at frequencies MSIP nitric oxide accounts for the inflow of arterial blood microcircular channel, increasing the number of functional capillaries and normalization of perfusion.

Studies indicate that male rats in a state of prolonged stress and subjected to a course of radiation at frequencies of terahertz waves MSIP nitric oxide 150.176 - 150.664 GHz is a restoration of perfusion index (M), while it is not statistically significantly different from the level of the control group. The animals in this group showed a statistically significant compared to the group of animals in a state of prolonged stress and irradiated terahertz waves, the increase in the standard deviation and the coefficient of variation of perfusion. It should be noted that the value of the standard deviation and coefficient of variation perfusion reaches the level values of the control group, indicating incomplete recovery flow modulation and the mechanisms of its regulation.

These amplitude and frequency analysis of LDF grams show that male rats subjected to irradiation with electromagnetic waves rate at frequencies MSIP nitric oxide in extended immmoblizative stress occurs statistically significant increase in amplitude and endothelial vasomotor oscillations compared to a group of animals in a state of prolonged immobilization stress, do not suffer TGCH-effects. This represents an increase vozodilative activity of endothelium (basal activation of nitric oxide production) and a decrease in peripheral resistance. Also, there is an increase of respiratory and pulse amplitudes (cardiac) oscillations, which indicates an increase in blood flow in the microcirculation. However, all indicators amplitudnochastotnogo analysis of LDF-gram animals in this group was significantly lower than the control group. This indicates a partial restoration of basal nitric oxide production, peripheral resistance and blood flow in the microcirculation.

When analyzing the results of termoprobe of skin found that under the action of waves TGCH male rats in a state of prolonged immobilization stress completely normalized initial and maximum perfusion and perfusion after initial restoration of blood flow. Scaling maximum perfusion in these animals suggests restoring secretion of nitric oxide induced endothelial capillary reserve power animals of this group is significant variability , which covers substantially quartile range as the control group and the group of animals in a state of prolonged stress is not exposed terahertz waves. This indicates that the male rats in a state of prolonged immobilization stress subjected to irradiation rate terahertz waves is partially retained strain generating microvessels, that is initially operates minimal number of capillaries than the animals of the control group.

Continuous and fractional modes of the electromagnetic radiation at a frequency of MSIP 129.0 GHz atmospheric oxygen contributes to the restoration of impaired aggregation activity of male rats in a state of acute stress immoblization. The degree of recovery of impaired platelet function depends on the mode of exposure.

TGCH-efficiency continuous impact on male rats with acute immobilization stress on frequency 129.0 GHz atmospheric oxygen depends on the exposure time. The most effective in restoring impaired platelet aggregation activity of the males is continuous irradiation for 5 minutes, 15- and 30-minute irradiation, however, inhibit platelet aggregation.

The effectiveness of the fractional mode TGCH-effects on male rats with acute immobilization stress on the frequency 129.0 GHz atmospheric oxygen depends on the exposure. The most effective in restoring the impaired platelet aggregation males is fractional mode total exposure of 5 minutes, which partially restores the impaired platelet aggregation, and fractional exposure mode total duration of 30 minutes causes the inhibition of platelet aggregation process.

Comparative analysis of the effect of electromagnetic radiation in the terahertz frequency MSIP atmospheric oxygen 129.0 GHz for the impaired functional activity of platelets in male rats in a state of acute immobilization stress showed that the continuous mode TGCH-efficient fractional irradiation.

For continuous mode of irradiation for 5 minutes more effectively recovers main indicators of platelet aggregation than the fractional mode with the same exposure.

Continuous TGCH impact on the frequency of atmospheric oxygen at 129.0 GHz females Dioestrus phase of the estrous cycle by an acute stress immoblization for 5 minutes only partially contributes to the restoration of impaired platelet aggregation, irradiation for 15 minutes, completely restores the violation of platelet aggregation, 30-minute exposure inhibits the functional activity of platelets.

Comparative analysis of sex differences in reactions to terahertz radiation at a frequency of 129.0 GHz atmosphere oxygen by an acute immobilization stress showed that male albino rats are more sensitive than females in the phase of the estrous cycle to Dioestrus terahertz radiation at a frequency of MSIP atmospheric oxygen 129.0 GHz.

The previous acute immobilization stress TGCH-irradiation on the frequency of atmospheric oxygen MSIP 129.0 GHz is capable of preventing the development of stress disorders platelet function in male rats. A more pronounced effect has 15- and 30-minute mode, while in continuous mode is 5 minutes of exposure.

The irradiation of white male rats in a state of stress immoblizative TGCH-waves at frequencies MSIP NO resulted in a statistically significant increase of plasma nitrite concentration of male rats in a state of acute stress immoblizative not subjected TGCH-effects. Stable and low concentrations of nitrite in the blood plasma of male rats in a state of acute immobilization stress on the background of L-Name indicates that the realization of the effect TGCH-wave frequency is specified with the participation of NO-synthase components of the cycle NO.

Increased activity of NO, leading to increased concentrations of nitrite in plasma under the influence of terahertz electromagnetic radiation makes it possible to develop new and effective drug-free correction techniques to reduce the endogenous synthesis of NO, which can be used in the treatment of diseases of the cardiovascular system involving production deficit NO (coronary artery disease, hypertension, etc.) studies suggest that exposure to electromagnetic waves in the terahertz frequency range of molecular absorption and emission of nitrogen oxide 150.176 - 150.664 GHz male rats before immobilization can prevent specific to acute immobilization stress perfusion abnormalities microcircular channel skin. Terahertz waves specified frequency effectively prevents the deterioration of the average perfusion and microcirculation disturbance modulation mechanisms. In a comparative analysis of the two exposure modes terahertz waves detected that the average efficiency of effects on microvascular perfusion prior to immobilization when the application TGCH irradiation at frequencies of nitric oxide and terahertz waves impact against developing the same immobilization stress. It was found that animals with prolonged immobilization stress are a persistent decrease in perfusion and blood flow in the modulation of inhibition microcirculation of the skin. These changes are more pronounced in comparison with acute immobilization stress. It is shown that a course of radiation at frequencies of terahertz waves MSIP NO: 150.176 - 150.664 GHz after each immobilization of male rats by prolonged stress can partially restore perfusion violation of the circular bed of the skin. This manifests the full normalization of the mean perfusion (perfusion index). Course terahertz wave irradiation of said frequency modulation is only partially normalizes blood flow in the skin microcircular vein of male rats during prolonged immobilization stress, as evidenced by partial restoration of perfusion standard deviation and coefficient of variation.

Terahertz radiation at frequencies MSIP 150.176 - 150.664 GHz partially or completely inhibits the intensity of lipoperosidation reduces the phenomenon autointaksication and cytolysis reactivates all links in the antioxidant and anti-radical protection, depending on the experimental conditions and time of exposure.

The irradiation of animals with acute immobilization stress the most effective was the 30-minute exposure mode. In this case, there was no statistically significant difference from the control group of the studied parameters characterizing the reaction of lipid peroxidation and antioxidant activity of blood. Effect of 5- and 15-minute frequencies of terahertz radiation on the MSIP nitric oxide 150.176 - 150.664 GHz by an acute stress did not cause immoblizative positive dynamics in terms of the antioxidant system and lipid peroxidation. So, at the 15 minute exposure mode terahertz waves was no change in all of the studied parameters in comparison with the group of animals in a state of acute stress. At the 15-minute exposure occurred only a partial restoration of parameters describing the processes of lipid peroxidation and antioxidant capacity of the blood of animals.

The most effective mode of terahertz radiation of animals on the background of chronic stress was also a 30- minute regime daily for 5 days of exposure. Against the background of this mode of action is completely normalized the activity of the antioxidant system of blood and there was a decrease to the level of intact animals, concentrations of toxic intermediate products of lipid peroxidation. At the 5-minute exposure terahertz waves at frequencies of nitric oxide 150.176 - 150.664 GHz not recovering indicators of antioxidant system and lipid peroxidation, as evidenced by the lack of a statistically significant difference compared with the group of stressed animals. At the 15-minute daily regime for 5 days of exposure of animals by chronic stress is a partial recovery of the parameters studied the antioxidant system and a partial normalization of intermediate concentrations of free radical oxidation products.

Based on the study it can be concluded about the positive impact of terahertz radiation at frequencies of 150.176 - 150.664 GHz on the changed parameters of the metabolic status of the experimental animals are in a state of acute stress immobilizption. The most effective in normalizing the study parameters is a 30-minute exposure mode. At the 15-minute exposure mode, there is a positive effect on the indicators of metabolic processes and metabolism (normal triglyceride concentration, the activity of AST and glutathione-S-transferase), and at 5- minute mode is absent.

In conditions of prolonged immobilization shown that in parallel with the action of the stress agent daily for 5 days of irradiation of terahertz waves at frequencies of molecular spectrum of emission and absorption of nitric oxide 150.176 - 150.664 GHz prevents the development of stress-dependent changes in the metabolic status of male rats which indicates the theoretical possibility of correction of these changes in this type of exposure. Thus, the daily for 5 days, the application of terahertz radiation at frequencies of nitric oxide 150.176 - 150.664 GHz to 30 minutes there is complete recovery of all the studied parameters that characterize the metabolic activity of the body's internal environment of experimental animals. The daily for 5 days exposure range of said electromagnetic waves for 15 minutes against prolonged stress is observed only a partial normalization of the metabolic status (statistically significantly reduced triglyceride levels). The impact of terahertz radiation at frequencies of nitric oxide 150.176 - 150.664 GHz to 5 minutes for 5 days in the stressed animals did not cause statistically significant changes in the metabolic status of the studied parameters. On the basis of experimental studies of the "Central Institute of scientific and research instrumentation" (Saratov g) was established EHF therapy "Orbit" in three versions YaKU L 941 526 001 941 526 001-01 Yakulov And Yakulov 941 526 001-02, for which the Federal Service on surveillance in Healthcare and Social Development issued a license number 99-03-002043 on June 7, 2010 and recorded license NQ FSR 2009/05497 on the production, sale and use of the territory of the Russian Federation These devices have been clinically tested in clinics, Saratov State Medical University, the Moscow Scientific Research Institute of Oncology PA Herzen Moscow city hospital number 15. Published guidelines "Using terahertz therapy in clinical practice" (Saratov, 2011). According to the results of experimental studies, received 24 patents for invention, and 2- the utility model, defended his doctoral dissertation (E. V. Andronov, 2008; O. N. Antipov, 2011), ten PhD theses (M. Bullock, 2001; L. I. Malinova, 2002; N. V. Mamontova, 2006; O. I. Pomoshnikova, 2006; A. N. Ivanov, 2007; S. V. Suhova, 2009; T. S. Velykanova, 2011; T. S. Kiriyazi, 2011; S. V. Svistunov, 2011) prepared for the protection of two doctoral dissertations (A. N. Ivanov; A. A. Tsymbal).

REFERENCES

1. Golant, M.B., Bryukhova, A.K., Dvadtsatova, E.A., et al. (1983) The Possibility of Regulating the Activity of Microorganisms When Exposed to Electromagnetic Waves MM Range. In: Devyatkova, N.D., Ed., The Effects of Non-Sexual Effects of

Millimeter-Wave Radiation on Biological Objects, M. IRE RAS, 115-122.

2. Devyatkov, N.D. and Betsky, O.V. (1985) Peculiarities of Interaction of Low-Intensity Millimeter Radiation with Biological Objects—The Use of MM-Radiation of Low Intensity in Biology and Medicine. Scientific Digest, 108-117.

3. Golant, M.B. (1986) Effect of Low-Intensity Monochromatic Electromagnetic Radiation of Millimeter Range on Biological. Biophysics, 31, 6-11.

4. Devyatkov, N.D., Golant, M.B. and Betsky, O.V. (1994) Features of Biomedical Applications of Millimeter Wave. M, IRE, RAS, 160.

5. Kiruchuk, V.F. Golovacheva, T.V. and Chizh, A.G. (1999) EHF-Therap. Saratov State Medical University Publishing House, Saratov, 360.

6. Kirichuk, V.F., Andronov, E.V., Mayborodin, A.V., et al. (2004) Effect of Electromagnetic Radiation at a Frequency Range of Molecular Absorption and Emission of Nitrogen Oxide (240 GHz) on the Aggregation Activity of Platelets in Patients with Unstable Angina. Angiology and Vascular Surgery, 10, 12.

Chapter 5

Explosions and Seismic Phenomena Based on Exciting Acoustic-Electromagnetic Waves

Svetlana Koshevaya[1*], Vladimir Grimalsky[1], Gustavo Urquiza[1], Margarita Tecpoyotl[1], Anatoliy Kotsarenko[2], Vsevolod Yutsis[3], Nikolay Makarets[4]

[1]*CIICAp, Autonomous University of State Morelos (UAEM), Cuernavaca, Mexico;* [*]*Corresponding Author: svetlana@uaem.mx*

[2]*Center of Geosciences, National Autonomous University of Mexico (UNAM), Juriquilla, Mexico*

[3]*Faculty of Earth Sciences, Autonomous University of Nuevo Leon (UANL), Linares, Mexico*

[4]*Physical Faculty, Taras Shevchenko National University, Kyiv, Ukraine*

ABSTRACT

During earthquakes and strong underground explosions it is possible to observe two different effects. The first one is connected with the destruction of media, and this causes acoustic and later hybrid acoustic-electromagnetic waves in an epicenter in the atmosphere and in the ionosphere. Another one is connected with cracks in crystals of rocks, which seems more interesting, because it is possible to recognize the natural earthquakes and industrial explosions. In the first effects due to nonlinear elastic modules the acoustic waves move through the lithosphere and transform their spectra from VLF (very low frequencies ~ 1 - 10 kHz) at a depth of about 30 km into the lower part of ELF (extremely low frequencies ~ 3 Hz - 1 kHz) on the Earth's surface, then they pass the atmosphere and penetrate into the ionosphere. During the nonlinear acoustic passage through the atmosphere and the ionosphere, the spectrum transforms from ELF into ULF (ultra low frequencies < 1 Hz) range. In this review article the classification of spectrum and analysis of two cases of the destruction of rocks in the lithosphere are presented. The rocks possess piezoelectric and piezomagnetic properties. In this case, the electromagnetic emission is excited by the fracturing in plates of crystals. The difference of emission from piezoelectric and magnetic plates in cases of industrial explosions and natural seismic events including volcanic phenomena gives a possibility to analyze the method of its identification. The consideration is based on the model of the plate of a finite size with a uniformly moving crack.

Keywords: Underground Explosion; Nonlinear Passing of Acoustic Waves; Piezoelectric and Magnetic Plates; Difference between Natural Seismic; Volcanic and Industrial Events

1. INTRODUCTION

It is very important to investigate the mechanisms of the energy flows from the lithosphere into the atmosphere and the ionosphere caused by natural hazards (seismic and volcano activity, tsunami etc.). All mechanisms have different precursors [1,2] due three basic channels of the lithosphere-ionosphere connection, namely, electromagnetic, geochemical, and acoustic ones [3].

The acoustic channel of the lithosphere-ionosphere coupling seems quite effective. It is due to atmospheric acoustic waves excited by fluctuations of a terrestrial surface [4]. This channel is illustrated in different phenomena [5] like excitation of plasma waves and periodic structure in the ionosphere, the increase of transparency for radiowaves, linear and nonlinear generation of magnetic perturbations, and oscillations of Eand F-layers in the ionosphere caused by acoustic and acoustic-gravity waves, nonlinear transformation of spectra of waves in the atmosphere, the ionosphere, and the lithosphere [5- 10]. The last case is analyzed below in two cases: strong seismic explosions in simple media and in case of cracks in a crystalline plate [11].

2. MODEL AND EQUATIONS

The geometry of the excitation of waves by underground fracturing for two cases is in Figure 1. One case is connected with the destruction of media, and this causes acoustic and later hybrid acoustic-electromagnetic waves in epicenters in the atmosphere and in the ionosphere. Another case is connected with cracks in crystals of rocks. Both cases are illustrated in the same Figure 1.

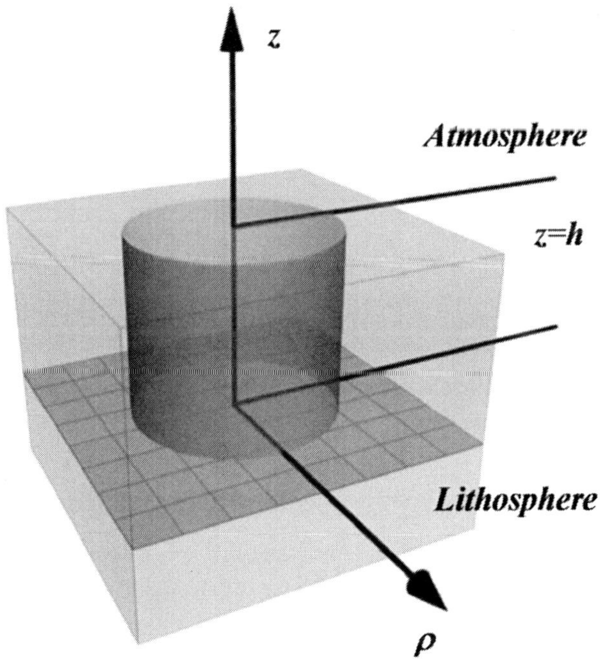

Figure 1. Geometry of the nonlinear passage of acoustic waves caused by underground fracturing through the lithosphere into the ionosphere.

The destruction and the excitation of waves including the case of cracks occur in an underground place in the depth h. After that the waves move to the Earth's surface and into the atmosphere and the ionosphere.

It is a cylindrical surface around the place which is the source of waves. After that, the seismic burst-like envelope of a finite transverse scale begins their passing through the Earth to the lithosphere, the atmosphere, and the ionosphere. Their passing through the Earth is nonlinear for two cases mentioned in the introduction.

In the case of cracks in rocks it is necessary to analyze the piezo and magnetic crystals [12], and the whole process has been considered of a finite size with a uniformly moving crack. Its linear front is parallel to the optical axis of the crystal. The tip of the crack instantly begins the uniform motion at a time moment t = 0 and causes the electromagnetic emission at a moment t = T > 0.

3. NONLINEAR PASSING OF ACOUSTIC WAVES

The first case is the seismic destruction in the lithosphere at the depth h (see Figure 1), it is described by the elasticity theory in the geometry of cylindrical surface around the place in which there is the source of underground seismic explosion (or deformation of a plate). The elasticity theory for the case of the Earth's crust as an isotropic uniform medium has the following the Newton's equation for the mechanic displacement $\boldsymbol{U}$:

$$\frac{\partial^2 \boldsymbol{U}}{\partial t^2} = \nabla\left\{\left(S_l^2 - S_t^2\right)\cdot \mathrm{div}\boldsymbol{U}\right\} + S_t^2 \cdot \Delta \boldsymbol{U} + \Gamma(z)\frac{\partial}{\partial t}\Delta \boldsymbol{U} + B(z)\nabla\left\{(\nu \boldsymbol{U})^2\right\} \tag{1}$$

The coefficients G(z) and B(z) describe the viscosity and nonlinearity of the elastic isotropic uniform medium in cylindrical system of coordinates for the acoustic waves with longitudinal and transverse velocities $S_{l,t}$ respectively.

The acoustic propagation in the atmosphere and the ionosphere is described by the following set of hydrodynamic equations:

$$\rho_0\left(\frac{\partial v_z}{\partial t}+v_z\frac{\partial v_z}{\partial z}\right)+\rho'\frac{\partial v_z}{\partial t}$$

$$\approx-\frac{\partial p'}{\partial z}+\frac{\partial}{\partial z}\left(\rho_0 v(z)\frac{\partial v_z}{\partial z}\right)-\rho' g;$$

$$\rho_0\frac{\partial \mathbf{v}_\perp}{\partial t}\approx-\nabla_\perp p'; \tag{2}$$

$$\frac{\partial\rho'}{\partial t}+\frac{\partial}{\partial z}(\rho_0 v_z)+\rho_0\mathrm{div}\mathbf{v}_\perp+\frac{\partial}{\partial z}(\rho' v_z)\approx 0;$$

$$\frac{\partial p'}{\partial t}-\rho_0 g v_z+v_z\frac{\partial p'}{\partial z}$$

$$\approx-c_s^{\,2}\left(\rho_0\frac{\partial v_z}{\partial z}++\rho_0\mathrm{div}\mathbf{v}_\perp+\gamma p'\frac{\partial v_z}{\partial z}\right)$$

Here $\left(v_x,v_y,v_z\right)$ are the components of the velocity of air $v_\perp\equiv\left(v_x,v_y\right)$; r, p are its total density and pressure; g is free fall acceleration; r = r_0 + r', p = p_0 + p'; where r_0, p_0 are the stationary values of atmospheric density and pressure $\left(\rho_0,p_0\sim\exp(-z/H)\right)$; and r', p' are alternating parts. The parameter H » 8 km is the average height of the atmosphere.

Additionally, this model uses the adiabatic equation for pressure with the adiabatic constant γ, and $v(z)\approx v(0)\exp(-z/H)$ is the kinematical viscosity of air. It is assumed that nonlinearity is moderate and only quadratic nonlinear terms are preserved in Eq.2. Note that all mechanisms of nonlinearity are essential. Also it is supposed that the acoustic wave moves preferentially vertically upwards, and inequalities $|v_{x,y}|$ << v_z are valid. In a linear approximation, when viscosity is neglecting, from Eq.2 one can see that for the acoustic wave $v(z,t) \approx \exp(-z/2H) \cdot \exp(i(\Omega t - Kz))$, and the energy flux is $\rho_0(z)|V(z,t)|^2 = \text{const}$.

We take into account the following dispersion law for vertically propagating atmospheric acoustic waves (AW):

$$\Omega^2 = \Omega_a^2 + K^2 c_s^2 \tag{3}$$

Here $\Omega_a = c_s/2H$ is the cut-off frequency, and c_s = $(\gamma p_0/\rho_0)^{1/2}$ is the speed of sound. Because of the fact that the attenuation of waves caused by the viscosity strongly grows with an increase of the frequency (as $\sim\Omega^2$), it is possible to expect that the effect of increasing atmospheric acoustic waves is essential only at frequencies W £ 10 s^{-1}. The observations confirmed this conclusion [1]. Within the lithosphere the vertical z-component of the acoustic wave is only demonstrated, the transverse profile is really smooth, and in modeling it is possible to neglect by the transverse acoustic components.

For the z-component the following equation is valid

$$\frac{\partial U_z}{\partial z} - \frac{\Gamma(z)}{2S_l^3} \cdot \frac{\partial^2 U_z}{\partial \eta^2} + \frac{B}{2S_l^4} \cdot \left(\frac{\partial U_z}{\partial \eta}\right)^2 - \frac{S_l}{2} \cdot \int_{-\infty}^{\eta} U_z(\eta') \cdot d\eta' = o, \tag{4}$$

This is the equation of the Khokhlov-Zabolotskaya [11]. For the case of the atmosphere and the ionosphere it is necessary to introduce a new dependent variable V = $v_z \exp(-z/2H)$, and the independent ones z, $\eta = t - t/c_s$, x [11]:

$$\frac{\partial}{\partial \eta}\left(\frac{\partial V}{\partial z}-\frac{v(0)e^{\frac{z}{H}}}{2c_s^{\ 3}}\frac{\partial^2 V}{\partial \eta^2}-\frac{1+\gamma}{2}\frac{e^{\frac{z}{2H}}}{2c_s^{\ 2}}\frac{\partial}{\partial \eta}\left(V^2\right)\right)$$

$$=\frac{c_s}{2}\frac{\partial^2 V}{\partial x^2} \qquad (5)$$

The evolution of low frequency part in the lithosphere described by Eqs.4 and 5 is shown in Figure 2. The initial burst is described in the following manner:

$$A_1(z=0,p,t)$$
$$=\left(A_{01}\exp(i\Delta\omega\cdot t)+A_{02}\exp(-i\Delta\omega\cdot t)\right)\exp\left(-(\rho/\rho_0)^2\right)$$

The normalizing for amplitudes of AW is A_0 = 0.01 cm^{-1}, ratio of nonlinear and linear elastic modules is $c_{111}/c_{11}=10$, a dissipation coefficient for the frequency $\omega_0=5\times10^4\mathrm{s}^{-1}$ is taken 0.25 km^{-1}.

The carrier frequency for ELF wave is chosen as f_1 = $\omega_1/2\pi\approx 80$ Hz.

The input ELF wave is burst-like, i.e., it is a wave packet with an ELF carrier frequency ω_0 ~ 10 - 500 s^{-1}; its envelope (modulation) frequency W ~ 0.1 - 5 s^{-1} is in the ULF frequency range:

$$V(w=1,t,\rho)=V_0\cos(\omega_0 t)\cdot\cos(\Omega t)\cdot\exp\left(-(\rho/\rho_0)^2\right).$$

The value of ρ_0 determines a transverse scale of the initial pulse. The Eq.5 is solved numerically by using spectral methods.

After an excitation at the Earth's surface, the ELF wave is subject to nonlinearity that leads to a generation of higher harmonics and down-conversion, namely, the increasing of the ULF component. More than 100 harmonics are excited. The growth of higher harmonics leads to the formation of a saw-tooth-like structure. Such a structure dissipates due to viscosity. The ULF part of spectrum is not subject to dissipation, on the contrary, it increases due to this nonlinear interaction. When the

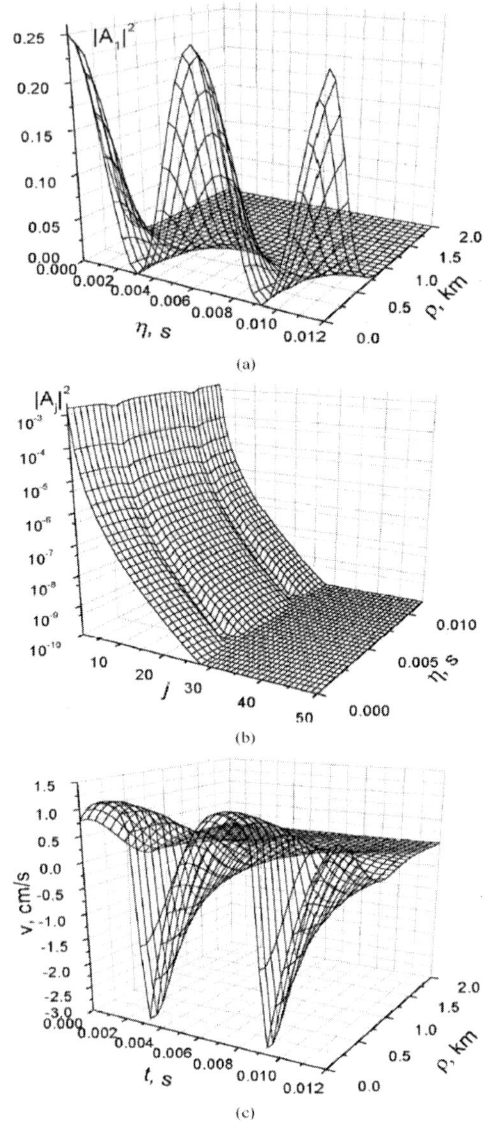

Figure 2. The nonlinear frequency conversion of AW (VLF AW – ELF AW) under passing through the lithosphere. (a) is initial distribution $|A_1(t,\rho)|^2$ of the first harmonic; (b) is a distribution of $|A_j(t,\rho=0)|^2$ for harmonics (j = 1,2,⋯), when the greatest ratio $|A_2/A_1|^2$ is observed; (c) is output distribution of the velocity of the ELF AW v(t,r) on the Earth's surface. Here $w_0 = 5 \times 10^4\ s^{-1}$, $w_{ELF} = 1000\ s^{-1}$; $r_0 = 0.5$ km.

transverse scale is ρ_0 ~ 0.3 - 10 km, the diffraction is not important for the ELF wave but it can decrease the peak amplitude of the ULF part. The results of the numerical simulations are given in Figure 3.

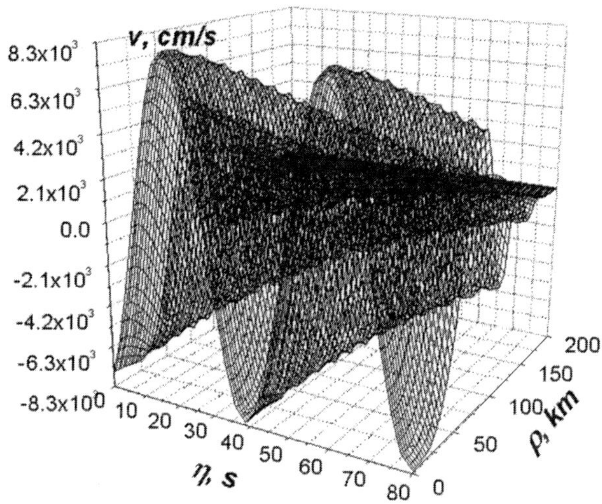

(a)

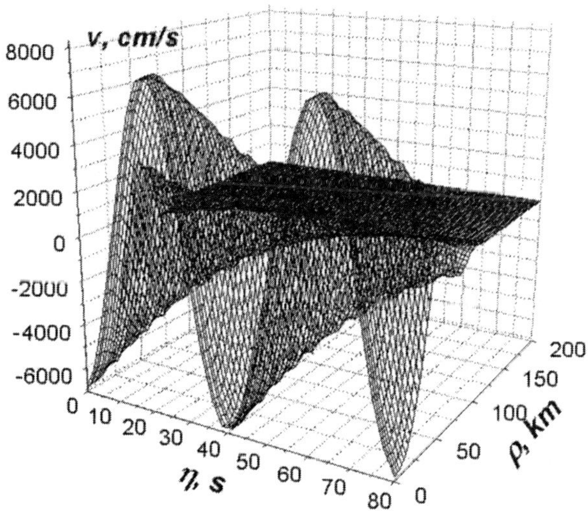

(b)

Figure 3. Nonlinear frequency conversion of AW. The excitation of ELF AW with the amplitude of V_0 = 30 cm/s. Here (a) is the distribution of velocity of the ULF AGW v(t,r) at the altitude z = 200 km; (b) is the distribution of velocity of the ULF AGW v(t,r) at an altitude z = 300 km. Here $w_0 = 30\ s^{-1}$, $W = 0.075\ s^{-1}$; r_0 = 25 km.

The efficiency of the nonlinear acoustic transformation is quite high. Within the framework of the specified model, the ULF acoustic-gravity waves (AGW) at frequencies 0.05 - 0.5 s^{-1} may reach the heights z = 200 - 300 km, where the maximum values of the density of free carriers of the ionosphere plasma occur.

In this analysis it is possible to take the classification of the spectrum. The underground fracturing causes VLF (very low frequencies ~ 1 - 10 kHz) at the depth of about 30 km, this initial spectrum changes during the passage into ELF (extremely low frequencies ~3 Hz - 1 kHz) on the Earth's surface, then AW pass the atmosphere and penetrate into the ionosphere. During the nonlinear acoustic passage through the atmosphere and the ionosphere, the spectrum transforms from ELF into ULF (ultra low frequencies < 1 Hz) range.

4. CRACKING OF PIEZOELECTRIC AND MAGNETIC PLATES

The second case of explosion in rocks is connected with exciting of cracks. It is supposed that a piezo and magnetic crystal surrounds by a plate possessing similar mechanical and dielectric properties. It is demonstrated that time-dependent mechanical stresses, which exist in the vicinity of the apex of the crack, are sources of EM waves, see below.

Additionally the comparison of the EM emission (EME) due to piezoeffect and piezomagnetism results in the different conclusions. The EME is investigated for the piezoelectric crystals of the point symmetry group 32 and for the piezomagnetic crystals of Fd3m space group symmetry in the first approximation for piezocoefficients and in the zero approximation for deviations of elastic tensor of the crystal from one for an isotropic medium with similar mechanical properties. It is obtained that the magnetic field of the emission has lots of maxima as a function of its frequency; these maxima are determined by the crystal size and the velocity of the crack and are located in LF and ELF ranges of the spectrum. It is found also that a source of the emission is equivalent to a timedependent magnetic dipole for both piezoelectric and piezomagnetic crystals, and the values of the magnetization vector and corresponding magnetization current have been estimated. Then the equations of the motion of the medium and equations of the EM field, written in the crystallographic frame of the piezocrystal are:

$$\rho\frac{\partial^2 U_i}{\partial t^2}=\partial_j\sigma_{ij};$$
$$\nabla\times\boldsymbol{H}=\frac{\partial\boldsymbol{D}}{\partial t};\nabla\times\boldsymbol{E}=-\frac{\partial\boldsymbol{B}}{\partial t};$$
$$\nabla\cdot\boldsymbol{D}=0;\nabla\cdot\boldsymbol{B}=0, \tag{6}$$

where r is the crystal density, U_i and s_{ij} are components of the displacement vector and mechanical stresses tensor, respectively, $¶_j$ denotes a partial derivative with respect to x_j; $\boldsymbol{E},\boldsymbol{H},\boldsymbol{D},\boldsymbol{B}$ are vectors of electric and magnetic field and corresponding inductions, respectively; c is the velocity of light. The constitutive equations for the piezoelectric crystal have the form:

$$\sigma_{ij}=c_{ijkl}U_{kl}-e_{kij}E_k;$$
$$D_i=\varepsilon_0\varepsilon_{ij}E_j+e_{ikl}U_{kl};$$
$$B_i=\mu_0\mu_{ij}H_j; \tag{7}$$

and for the piezomagnetic crystal:

$$\sigma_{ij}=c_{ijkl}U_{kl}-q_{kij}H_k;$$
$$D_i=\varepsilon_0\varepsilon_{ij}E_j;$$
$$B_i=\mu_0\mu_{ij}H_j+q_{ikl}U_{kl}; \tag{8}$$

The quantities U_{ij}, e_{ij}, and μ_{ij} in Eqs.6-8 are dimensionless, $U_{ij}=(\partial_i U_j+\partial_j U_i)/2$ are components of the strain tensor, e_0 and μ_0 are electric and magnetic constants, respectively, c_{ijkl}, e_{ikl}, q_{ijk}, μ_{ij}, e_{ij} are components of tensors of elastic, piezoelectric, piezomagnetic, magnetic, and dielectric constants, respectively.

Eqs.6-8 have been solved by using the Fourier transform on spatial coordinates within the infinite limits and the Laplace transform with respect to time t [12,13]. It is found that the non-stationary mechanical stresses in the vicinity of the moving tip of the crack have caused nonstationary polarization currents, which possess potential and vortical components inside a piezoelectric crystal and only vortical components inside a piezomagnetic one. The potential currents generate electric polarization current I_P, but the vortical ones just are magnetization currents I_M, or magnetic dipoles. In both cases namely the non-stationary magnetic dipole generates the electromagnetic radiation in the far zone and this dipole is created by the moving tip of the plate due to specific crystal symmetry related with the crystal piezoproperties (see Figure 4).

For estimations the material coefficients for piezoelectric quartz (in matrix representation from [14]) are used: $\varepsilon_1 \approx 4.6$, $e_{11} \approx 0.17$ C/m^2, $e_{14} \approx 0.04$ C/m^2. Now, it is necessary to use the estimations obtained in [11] for the magnetic moment and vortex current created by a single moving displacement crack through a crystal of quartz:

$$M(t) = \frac{1}{2}\iiint\limits_V \left[r \times j(r,t) \right] \mathrm{d}V \sim e_{14} l_0^{1/2} L^{5/2},$$

$$I \sim \frac{M}{L^2} \tag{9}$$

where L is the crystal size, and l_0 is the initial size of the crack, which is usually close to several interatomic distance. When we set L » 1 m, l_0 » 1 nm, the velocity of the motion of the crack is V_{cr} » 1.5 km/s, then we find M » 0.01··· 0.02 A·m^2 and corresponding polarization current I » 0.02 A.

It is possible to make the same for the magnetic moment and corresponding polarization current into the cracked piezomagnetic crystal $CoFe_2O_4$ by using the material coefficients from [13] q_{31} » 580, q_{33} » 700, and q_{15} » 550 N/A·m and approximations:

$$M_i \approx \max\left(\frac{q_{ikl}}{\mu_0} U_{kl} V\right) \approx \max\left(\frac{q_{ikl}}{\mu_0} l_0^{1/2} L^{5/2}\right),$$
$$I \approx \frac{M}{L^2} \qquad (10)$$

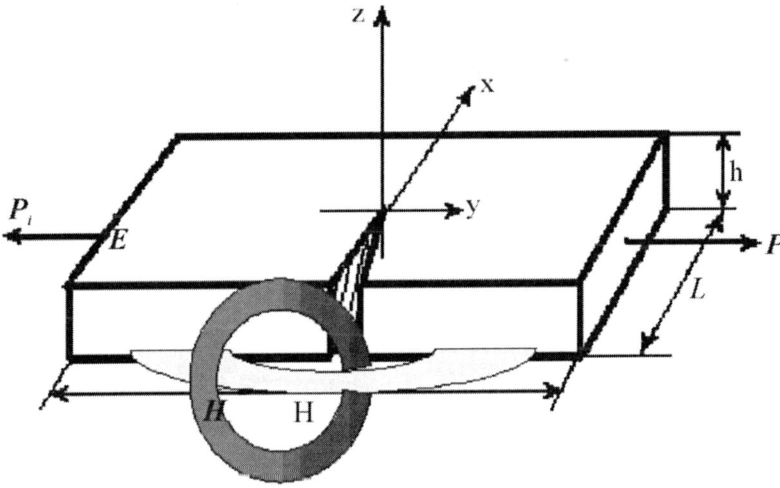

Figure 4. Moving crack and EM fields around it.

This gives M » 0.15 A·m^2 and I » 0.15 A. Therefore, the moving crack generates the vortex current inside piezoelectric materials, which is comparable with one inside the piezomagnetic ones. The pointed above magnetic moments, created by only one piezocrystal of the given size, at the distance r = 100 m generate the magnetic field with induction B_e » 1 ´ 10^{-6} nT for the piezoelectric case and B_m » 1 ´ 10^{-5} nT for the piezomagnetic one.

These fields are very small to detect a single crack only. But we suppose that inside the cracked rock massive with a characteristic size of about 100 × 100 × 100 m^3 there are 10^4 cracks. Hence, the magnetic field at the distance r = 100 m can potentially be enhanced up to 1 nT, when a great crack crosses such a rock massive and all piezocrystals generate correlated magnetic fields. In VLF range f ~ 1 kHz, the magnetic fields of pointed above magnitudes ~ 0.001··· 0.1 nT can be detectable. It is investigated the solutions of magnetic dipoles and magnetic field using an approximation for mechanical stresses both around of the crack, and for the wave of unloading.

In both cases it is necessary to find the effect that the frequency dependence of the magnetic field has two parts, created by unloading wave and by the tip of a crack, respectively.

The frequency dependencies of the magnetic field generated by the tip of the crack and wave of unloading are given in Figures 5 and 6. The dimensionless frequency is $\tilde{\omega} = L\omega/V_{cr}$.

One can see that the field created by the tip of the crack has several maxima at the dimensionless frequencies $\tilde{\omega}_{t,1} \approx 3$, $\tilde{\omega}_{t,2} \approx 9$, $\tilde{\omega}_{t,3} \approx 16$ whereas the one created due to unloading wave has a single one at $\tilde{\omega}_u$ » 1.5. The value of such a single maximum is dominating and exceeds the value of the first maximum due to the motion of the tip

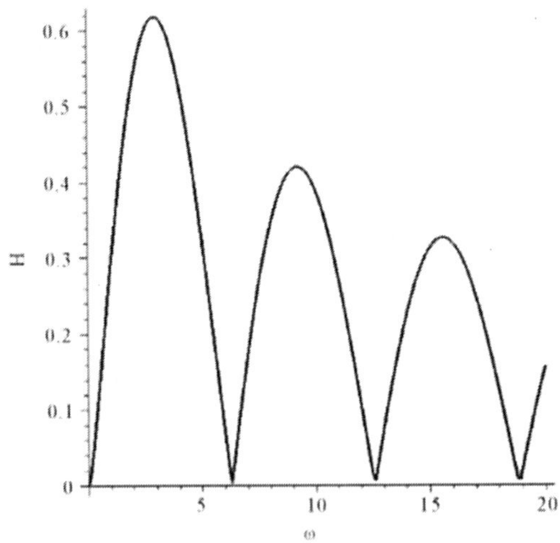

Figure 5. Frequency dependence of the magnetic field created by the tip of crack.

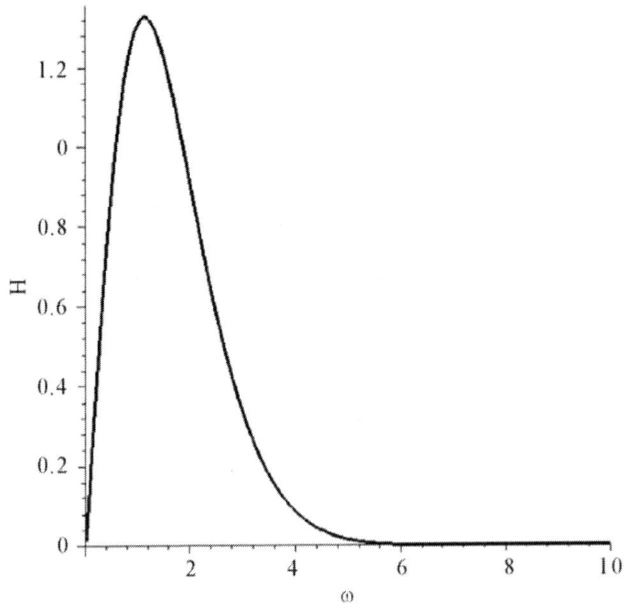

Figure 6. Frequency dependence of the magnetic field created by the unloading wave.

about 4 times. For the values L = 1 cm and V_{cr} = 4 km/s (a small crack) the corresponding frequencies are $f_{t,\perp} = \tilde{\omega}_{t,\perp} V_{cr} / 2\pi L \approx$ 190 kHz, $f_u = \tilde{\omega}_u V_{cr} / 2\pi L$ » 95 kHz. The fields at this frequency range dissipate strongly at small distances ~ 100 m. But for a large and slow crack with L » 1 m and V_{cr} » 1.5 km/s the corresponding frequencies will be $f_{t,\perp}$ » 720 Hz and f_u » 320 Hz. Therefore, the spectrum of the field can be very wide. Because the magnetic field is registered from numerous cracks of different sizes, it is possible to conclude that the resulting field lies in LF, VLF and ELF ranges of the spectrum: VLF (very low frequencies ~ 1 - 10 kHz) at the depth of about 30 km, ELF (extremely low frequencies, ~ 3 Hz - 1 kHz) on the Earth's surface, then they pass the atmosphere and penetrate into the ionosphere like ULF (ultra low frequencies <1 Hz) range.

Thus, obtained amplitude-frequency characteristics of the magnetic field generated by the fracturing of piezocrystals in rocks have several frequency scales. From these results one can explain some phenomena caused by seismic and volcanic activity. The main result of the calculations is that it is possible to see difference between seismic and industrial events by means of observation of the unloading wave only in the case of natural hazards. This opens a possibility to create the method of remote sensing and dividing the seismic and industrial events.

5. CONCLUSIONS

An analysis of the nonlinear transformation of the spectrum: VLF (very low frequencies ~ 1 - 10 kHz) at a depth of about 30 km , ELF (extremely low frequencies ~ 3 Hz - 1 kHz) on the Earth's surface, then they pass the atmosphere and penetrate into the ionosphere like ULF (ultra low frequencies < 1 Hz) range.

The nonlinear interaction of acoustic burst demonstrates additionally the importance of acoustic channel by the example of the nonlinear mechanism of the energy flow from the lithosphere into the ionosphere.

The destruction of rocks has been analyzed for cases of piezoelectric and magnetic crystals. In the both cases there exists the quasi-stationary magnetic dipole, which is the source of the magnetic field (and the source of electromagnetic radiation in the far zone, too). Such a dipole is created by the motion of the apex of the crack and depends on the specific crystalline symmetry that determines its piezoproperties. The field has several maxima as a function of the frequency. These maxima are determined by the crystal size and the velocity of the crack and are located in LF, VLF and ELF ranges of the spectrum. The values of the magnetization vector and the corresponding magnetization current are estimated. The value of magnetization current for a typical piezoelectric like quartz is smaller (but comparable) than one for a typical piezomagnetic material. The basic conclusion of the calculation is that it is possible to see the difference between natural seismic and industrial events by means of observation of the unloading wave. The unloading wave is absent in the case of non-natural events. This opens a possibility to create the method of remote sensing and dividing the seismic and industrial events.

6. ACKNOWLEDGEMENTS

Authors thank SEP (PROMEP)-CONACyT for the support of our work.

REFERENCES

1. Galperin, Yu.I., Gladyshev, V.A., Jorjio, N.V., Kovrazhkin, R.A., Lissakov, Yu.V., Maslov, V.D., Nikolaenko, L.M., Sagdeev, R.Z., Molchanov, O.A., Mogilevsky, M.M., Alperovich, L.S., Gokhberg, M.B., Ivanov, E.A., Pokhotelov, O.A., Beghin, C., Berthelier, J.J., Bosqued, J.-M., and Reme, H. (1985) Alfvén wave excited in the middle-latitude mag- netosphere by a large-scale acoustic wave propagating in lower ionosphere. Izvestiya, Physics of the Solid Earth, 21, 88-98.

2. Warwick, J.W. (1963) Radio astronomical and satellite studies of the atmosphere. North-Holland, Amsterdam.

3. Gokhberg, M., Morgounov, V. and Pokhotelov, O. (1995) Earthquake prediction: Seismo-electromagnetic phenomena. Gordon and Breach, London.

4. Tarantsev, A. and Birfeld, Ya. (1973) The phenomenon of the action of the Earth's seismicity on the ionosphere through the acoustic waves, September 1963. CNIIPII Publication, Moscow.

5. Kotsarenko, N., Enriquez, R.P. and Koshevaya, S. (1997) Excitation of plasma waves in the ionosphere caused by atmospheric acoustic waves. Astrophysics and Space Science, 246, 211-217.

6. Surkov, V.V., Molchanov, O.A. and Hayakawa, M. (2003) Pre-earthquake ULF electromagnetic perturbations as a result of inductive seismo-magnetic phenomena during microfracturing. Journal of Atmospheric and Solar-Terrestrial Physics, 65, 31-46.

7. Makarets, M.V., Koshevaya, S.V. and Gernets, A.A. (2002) Electromagnetic emission caused by the fracturing of piezoelectrics in rocks. Physica Scripta, 65, 268-272.

8. Khavroshkin, O.B., Tsyplakov, V.V. and Berezhnoi, A.A. (2001) Modification of the Moon's microwave radiation by the fall of lunar prospector. Doklady Earth Sciences, 376, 90-92.

9. Musatenko, S.I., Kurochka, E.V., Aslanyan, A.M., Borovkov, O.V., Gulyan, A.G., Ivchenko, V.N., Kotsarenko, A.N., Slipchenko A.S., Margorin O.K., and Nabatov, A.S. (1997) Fluctuations of Jupiter radio emission at 3.6 cm due to the impact of the fragments of Comet Shoemaker-Levy. Kinematika i Fizika Nebesnykh Tel (Kinematics and Physics of Celestial Bodies), 13, 64-73.

10. Berezhnoy, A.A., Bervalds, E., Khavroshkin, O.B., Kovalenko, A.V., Ozolins, G., Paupere, M., Smirnov and G.T., Tsyplakov, V.V. (2002) Radio observations of the Moon during activity periods of the Leonid and Lyrid meteor streams. Baltic Astronomy, 11, 507-527.

11. Rudenko, O. (1995) Nonlinear sawtooth-shaped waves. Physics-Uspekhi, 38, 965-989.

12. Koshevaya, S., Grimalsky, V., Makarets, N., Kotsarenko, A., Siquieros-Alatorre, J., Perez-Enriquez, R. and JuarezRomero, D. (2008) Electromagnetic emission from Magnetite plate cracking under seismic processes. Advances in Geosciences, 14, 25-28.

13. Gernets, A.A., Makarets, M.V., Koshevaya, S.V., Grimalsky, V.V., Juarez-Romero, D. and Kotsarenko, A.N. (2004) Electromagnetic emission caused by the fracturing of piezoelectric crystals with an arbitrarily oriented moving crack. Physics and Chemistry of the Earth, 29, 463- 472.

14. Vainshtein, V.K., Ed. (1981) Modern crystallography. Nauka, Moscow.

Chapter 6

Electromagnetic Waves Absorbing Characteristics of Composite Material Containing Carbonyl Iron Particles

Victor A. Zhuravlev, Valentin I. Suslyaev, Evgeny Yu. Korovin, Kyrill V. Dorozhkin

Department of Radioelectronics, National Research Tomsk State University, Tomsk, Russia

ABSTRACT

Results of measurements of permeability, permittivity and radar absorption properties of composites on basis of carbonyl iron particles R-10 brand are presented in this paper. The calculations and experimental studies have shown that in the super high frequency (SHF) and extremely high frequency (EHF) ranges on the basis of two-layer structures with different content of carbonyl iron particles can create a radar absorbing coatings with a reflectivity of less than −10 dB over a wide bandwidth from 3.1 to 17.1 GHz and from 27 to 37 GHz. Absorbing properties of composites are saved in terahertz frequency range from 250 to 525 GHz.

Keywords: Electromagnetic Wave Absorbers, Radar Absorbing Materials and Coatings, Composite Materials, Carbonyl Iron Powders, Spectrum of Permittivity and Permeability

1. INTRODUCTION

Active development of the microwave frequency range sharpened the need for magnetic materials absorbing electromagnetic radiation. Such materials are required to reduce harmful influence on biological objects, to provide electromagnetic compatibility of units and blocks of high-frequency devices, to build anechoic chambers and to protect information. Composite radar absorbing coating (RAC) which consists of a matrix of high-mole- cular polymer compounds and filler particles of ferromagnetic or ferrimagnetic materials are widely used [1] . Such materials are most effectively used as electromagnetic waves absorbers (EMWA) located on the metal surfaces.

The carbonyl iron powders (CIP) are used at works [2] -[10] as fillers for composite RAC in microwave range. Investigation of complex permeability ($\mu = \mu' - i\mu''$) and permittivity ($\varepsilon = \varepsilon' - i\varepsilon''$) spectra of composites with different concentration of CIP at frequencies up to 18 GHz shows that these materials have noticeable permeability and permittivity loses, and they are perspective as EMWA for microwave frequency range. The rather narrow operating frequency band is disadvantage of such RAC.

It is shown in [5] [6] that double-layer RAC is more broadband and has less reflection coefficient than single- layer RAC [1] -[3] [7] -[10] . The first layer of such RAC is an absorbing layer. It is located on the metal surface. The second layer is matching with free space. It has smaller values of the permeability and permittivity.

It is known [11] that frequency band of high magnetic loses caused by natural ferromagnetic resonance at presence of domain structure lies generally in the limits from $2\pi f_1 = \gamma H_a$ to $2\pi f_2 = \gamma(H_a + 4\pi M_S)$. Because of big value of saturation magnetization of iron $M_S \approx 1.7$ kGs the frequency band of high magnetic loses is from $f_1 \approx 1.4$ GHz up to $f_2 \approx 61$ GHz. In calculation it is accepted $\gamma/2\pi = 2.8$ GHz/kOe, anisotropy field $H_a \approx 0.5$ kOe. This estimation shows that composite materials on basis of CIP can have noticeable absorbing properties in wide frequency range, including SHF and EHF.

Results of investigation of permittivity and permeability spectra, absorbing properties of composite materials composed of CIP of R-10 brand in polyurethane matrix are presented in this paper. CIP of R-10 brand had onion like structure with average size of particles about 3.5 μm. Thickness of investigated flat samples was 1 mm , volume content of CIP 9% (sample 1) and 53% (sample 2).

2. THEORY

Theoretical analysis of radar absorbing properties of materials is usually carried out for the simplest case of normal incidence of a plane electromagnetic wave (EMW) on a magnetodielectric (MD) layer. Permeability (μ) and permittivity (e) of layer are taken as complex scalar quantities. For developing of radio protective materials and coatings located in free space layer of MD without metallization is considered. For reduction of radar signature a model of MD layer disposed on the metal surface is used. Formulas for calculating the complex reflection (R_T) and transmission (T_T) coefficients in the first case can be written in the form [12] :

$$R_{\mathrm{T}} = \frac{\rho\left(1-\exp(-2i\gamma d)\right)}{1-\rho^2\exp(-2i\gamma d)}, \quad T_{\mathrm{T}} = \frac{\left(1-\rho^2\right)\exp(-2i\gamma d)}{1-\rho^2\exp(-2i\gamma d)}. \tag{1}$$

Here: $\rho = (Z-1)/(Z+1)$ —reflection coefficient from the front edge of MD. $Z = \sqrt{\mu/\varepsilon}$ —wave impedance.

$\gamma = k_0\sqrt{\varepsilon\mu}$ —EMW propagation constant in MD. $k_0 = \omega/c$ —the wave number of free space. $\omega = 2\pi f$ —cir- cular frequency of the EMW. c—the speed of light. d—thickness of the layer. $i = \sqrt{-1}$.

Expression for the reflection coefficient (R_R) for located on the a metal plate layer of MD has the form [12] :

$$R_{\mathrm{R}} = \left[\rho - \exp(-2i\gamma d)\right] / \left[1 - \rho \exp(-2i\gamma d)\right] \quad . (2)$$

Commercial advantage has RAC with a small thickness and weight. Therefore it is interesting to analyze the expression (1) and (2) in the case of a thin layer and find out which material parameters are crucial to improve the efficiency of consumer qualities of such materials. Use the approximation of a electrically thin layer: $|\gamma d| \ll 1$. In the linear approximation formulas (1) are significantly simplified and coefficients R_T, T_T can be rewritten as:

$$T_{\mathrm{T}} \approx \left[1 + i(\mu + \varepsilon) k_0 d/2\right]^{-1}, \; R_{\mathrm{T}} \approx i T_{\mathrm{T}} (\mu - \varepsilon) k_0 d/2. \quad (3)$$

For the modules of the transmission and reflection coefficients are obtained the following expressions:

$$|T_{\mathrm{T}}| = 1 / \sqrt{1 + k_0 d(\mu'' + \varepsilon'')}, \; |R_{\mathrm{T}}| = |T_{\mathrm{T}}| \frac{k_0 d}{2} \sqrt{(\mu' - \varepsilon')^2 + (\mu'' - \varepsilon'')^2} \quad . (4)$$

According to formulas (3), (4) in the thin layer approximation a module of the transmission coefficients depends only on the imaginary parts of permittivity and permeability describing the magnetic and dielectric losses in the MD layer. The reflection coefficient vanishes when real and imaginary parts of the permeability and permittivity are equal. In this case there is equality of the wave impedances of the interface MD layer and free space. In real composites to achieve such equality it is impossible.

The expression (2) for the module of reflection coefficient located on a metal surface layer of MD at $|\gamma d| \ll 1$ can be written:

$$|R_{\mathrm{R}}| = \sqrt{(1 - 2k_0 d\mu'')/(1 + 2k_0 d\mu'')} \quad . (5)$$

The reflection coefficient depends only on the imaginary part of permeability MD in this case. The optimal thickness of the layer (d_{opt}) ensuring the absence of the reflected wave is equal to:

$$d_{\text{opt}} = 1/2k_0\mu'' = c/2\omega\mu''. \tag{6}$$

Equation (6) contains an important consequence for practical applications. To ensure the broad band RAC the imaginary part of permittivity of MD should vary inversely proportional to the frequency:

$$\mu'' \sim 1/\omega. \tag{7}$$

It should be noted that the formulas (4) and (5) provide an opportunity to solve another important experimental problem—to assess the values of the imaginary components μ'', ε'' from the measured values of $|R_{\text{R}}|$ and $|T_{\text{T}}|$.

Calculation of the reflection coefficient of multilayer RAC was carried out in [13] . The formula for the reflection coefficient from the interface between the n + 1 and n layers is follows:

$$R_{n+1} = \left(Z_{\text{in}}^{(n)} - Z_{n+1}\right)/\left(Z_{\text{in}}^{(n)} + Z_{n+1}\right). \tag{8}$$

Z_{n+1} —impedance of layer n + 1. $Z_{\text{in}}^{(n)}$ —input impedance of layer n. It is equal to

$$Z_{\text{in}}^{(n)} = \frac{1 + R_n \exp(-2i\gamma_n d_n)}{1 - R_n \exp(-2i\gamma_n d_n)} Z_n. \tag{9}$$

Equations (8) and (9) will be used later to calculate the reflection coefficient of the two-layer RAC located on the metal surface.

3. RESULTS AND DISCUSSIONS

3.1. Permeability and Permittivity Spectra of Composite

Spectra of permeability ($\mu = \mu' - i\mu''$) and permittivity ($\varepsilon = \varepsilon' - i\varepsilon''$) measured using vector network analyzer Agilent Technologies E8363B. Measurements were performed in the frequency range 10 MHz - 20 GHz. A coaxial waveguide (R/T method, [12]), an irregular microstrip resonator [14] [15] , and a set of rectangular multimode volume resonators [16] were used as measuring cells.

Since various experimental techniques have different systematic and random measurement errors, permeability and permittivity values measured by the corresponding methods at the ends of the examined frequency range differed. In this regard, the experimental magnetic spectra of the composites were processed by the method suggested in [17] [18] . To match the frequency dependences of the imaginary and real permeability components, the Kramers-Kronig relations and independent measurements of the initial static permeability (μ_0) of the materials were used. Here μ_0 values, equal to 10.6 and 3.1 for composites with CIP contents of 53% and 9%, respectively, were measured for toroidal samples at a frequency range 20 Hz to 2 MHz using precision LCR Meter Agilent E4980A.

Figure 1 and Figure 2 shows the magnetic permeability spectra of the composites.

It can be seen that a maximum in the frequency dependence of the imaginary part of permeability for both samples is localized near a frequency of 1 GHz and depends weakly on the filler concentration. The measured real and imaginary permeability components are correlated with the data presented in [7] -[10] for close values of CIP concentrations.

Our measurements of the permittivity spectra demonstrated that its variance was small in the examined frequency range, and the permittivity of samples remains virtually constant: $\varepsilon' = 28$, $\varepsilon'' = 4$ for the sample 2 and $\varepsilon' = 6.5$, $\varepsilon'' = 0.2$ for the sample 1.

3.2. SINGLE-LAYER EMWA SHF RANGE

It was shown above (see Equation (7)) that to develop thin broadband RAC on the metal surface, it is necessary that μ" varies inversely proportional to the frequency of the EMW. In the case of composite materials based on CIP, the imaginary permeability component decreases with frequency much more slowly.

Figure 3 and Figure 4 shows the frequency dependence of the reflection coefficients of the layers located on the metal surface calculated according to formula (2). The obtained above values of μ and e samples 1 and 2 are used in the calculations. Numbers near the curves-sample thicknesses in mm.

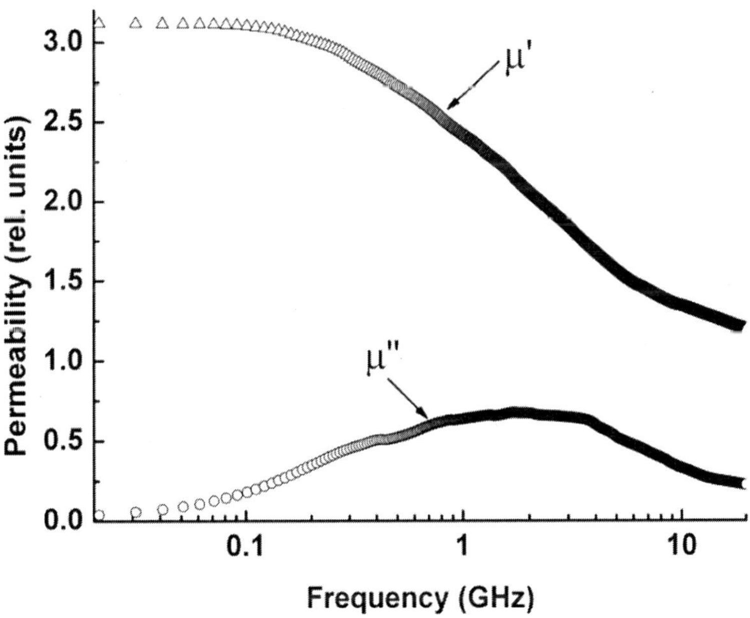

Figure 1. Permeability spectra of the sample 1 with CIP content of 9%.

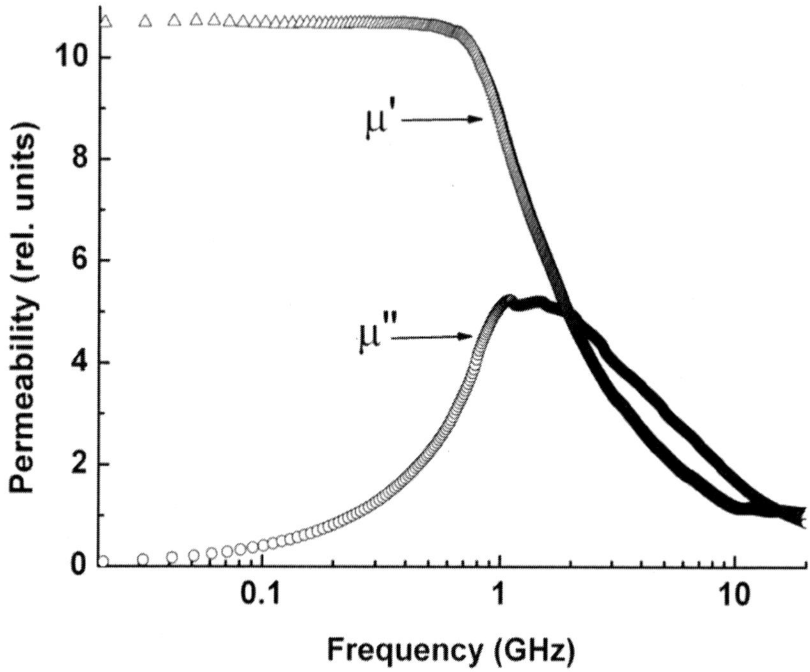

Figure 2. Permeability spectra of the sample 2 with CIP content of 53%.

Presented in Figure 3, Figure 4 results demonstrated that on the base of sample 1 and 2 it is difficult to make a single-layer RAC with the reflection coefficient less than −10 dB for the frequency band exceeding the octave.

3.3. Two-Layer EMWA SHF Range

We investigated a two-layer absorber design whose absorbing layer on metal surface was formed by the composite sample 2 (thick d_1) and whose matching layer was formed by the sample 1 (thick d_2) with small values of the permeability and permittivity. The calculation was performed according to the formulas (8) and (9). The simulation results presented on Figure 5 by solid lines.

Curve 1—the calculation for the absorbing layer thickness d_1 = 0.9 mm, matching layer d_2 = 2.3 mm. At such thicknesses reflection coefficient smaller −10 dB is observed in the frequency range from 3.1 to 17.1 GHz. Decrease in the thickness of the layers to d_1 = 0.78 mm and d_2 = 1.95 mm leads to a shift of the frequency band with a reflection coefficient ≤−10 dB in the high frequency region from 4 to 20 GHz (curve 2).

The points in Figure 5—experimental curve measured in the coaxial short-circuited line size of 7/3.05 mm. Measurements were performed on a two-layer construction with thickness d_1 = 1 mm (sample 2) and d_2 = 2 mm (sample 1). It can be seen that the deviation from of optimal thickness in the experimental RAC resulted in an increase of the reflection coefficient in the region of frequencies from 6 to 11 GHz. Thus, the use of bilayer RAC leads to substantial expansion of the band of operating frequencies as compared with monolayer.

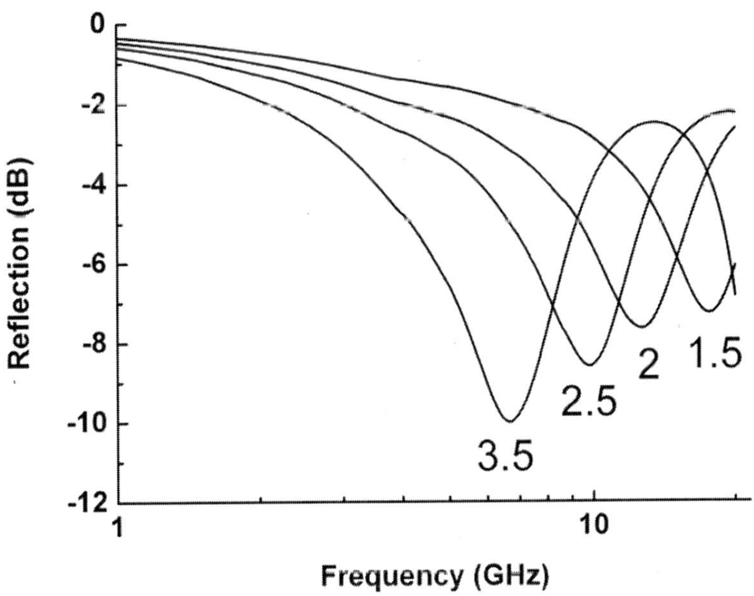

Figure 3. Reflection coefficients of the sample 1 calculated for different sample thicknesses. The numbers near the curves-RAC thickness in mm.

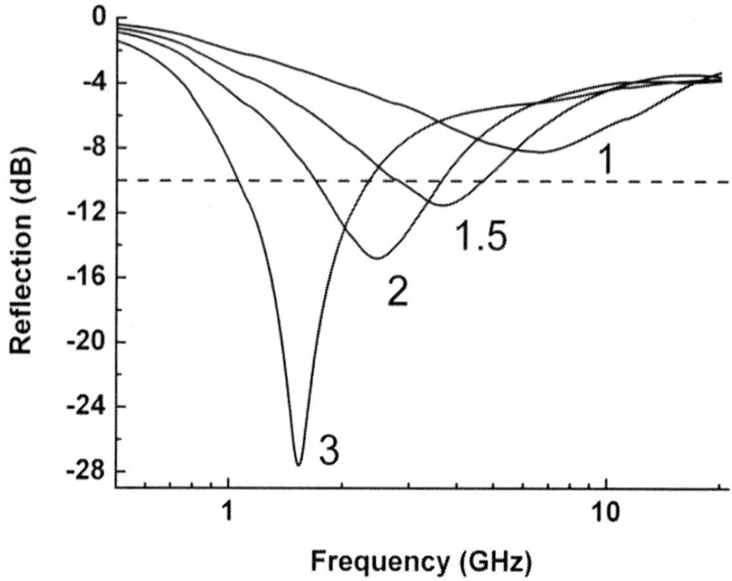

Figure 4. Reflection coefficients of the sample 2 calculated for different sample thicknesses. The numbers near the curves-RAC thickness in mm.

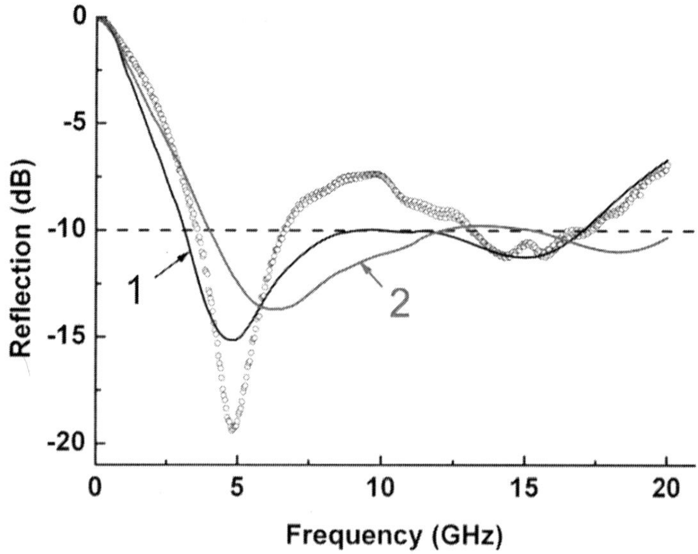

Figure 5. Reflectivity of two-layer RAC. Solid lines—calculation. Curve 1: $d_1 = 0.9$ mm, $d_2 = 2.3$ mm. Curve 2: $d_1 = 0.78$ mm, $d_2 = 1.95$ mm. Point —

an experimental results on a sample with thickness d_1 = 1 mm, d_2 = 2 mm.

3.4. Single-Layer and Two-Layer EMWA EHF Range

This section presents the results of experimental investigations of the absorbing properties of composites samples 1 and 2 for frequencies in the range 26 - 37 GHz. Measurements were performed using the rectangular horn antennas 5 cm long and 2 × l cm^2 antenna apertures. The SWR of the system of horn antennas without sample in the examined frequency range did not exceed 1.2.

Figure 6 shows the reflection coefficient of RAC located on the metal surface.

Here curve 1 is for sample 1 with thickness of d = 1 mm, curve 2 is for sample 2 (d = 1 mm), and curve 3 is for a two-layer design consisting of sample 1 (d_2 = 1 mm) and sample 2 (d_1 = 1 mm). From the figure it can be seen that the two-layer structure has the reflection coefficient much less than the single-layer structure, and a minimum of reflection is observed at frequencies greater than 37 GHz. As previously, sample 1 with the small content of CIP plays the role of a matching layer.

Figure 7 shows the results of investigations of the electromagnetic wave absorbers in the transmission/ref- lection mode. Values of the transmission and the reflection coefficients in decibels were measured. Then the measured values were recalculated into the transmission coefficient T and the reflection coefficient R on power in percent. The absorption coefficient was calculated from the formula A = 100 – R – T.

Figure 7(a) shows the measured values for sample 1. It can be seen that the reflection coefficient does not exceed 20%, and the transmission and absorption coefficients are close in values (about 40%). The corresponding values measured for sample 2 are shown in Figure 7(b). For this material with large CIP content, the reflection coefficient is somewhat higher. However, the absorbing properties of these RAC are almost twice greater. The transmission coefficient is about 1%. Figure

7(c) shows the measured values for the two-layer RAC design with 1-mm thick sample 1 used as a matching layer and sample 2 also with a thickness of 1 mm used as an absorbing layer. It can be seen that the two-layer absorber of electromagnetic waves in the given range of frequencies has the absorption coefficient greater than 90% and the reflection coefficient smaller than 10%. The transmission coefficient is small; it does not exceed 1%. Thus, good microwave properties of composites based on CIP observed in the EHF range. This is due to the wide frequency band in which there are losses on EFMR.

3.5. Study of the Samples 1 and 2 in the Terahertz Frequency Range

The electromagnetic response of samples 1 and 2 was also investigated in frequency range 150 - 525 GHz. The terahertz spectrometer STD-21 with backward-wave oscillators were used as a source of electromagnetic wave. We investigated the frequency dependence of transmission and reflection coefficients. The results of measuring complex permittivity in this frequency range are shown in Figure 8 and Figure 9.

It can be seen that the real and imaginary parts of permittivity vary slightly within a given range of frequencies. Determined from experiments values of permittivity correlate with the above performed estimates for e in SHF range.

Transmission, reflection and absorption coefficient on power for samples 1 and 2 are shown in Figure 10 and Figure 11.

Figure 10 and Figure 11 shows that the transmission coefficient for both samples is small and does not exceed 1%. For the sample with low content of CIP reflectance in the investigated frequency range does not exceed 10% and that of sample 2, it is about 30%. The remainder part of EMW power is absorbed. Thus, investigated composites have good absorbing properties in the terahertz frequency range.

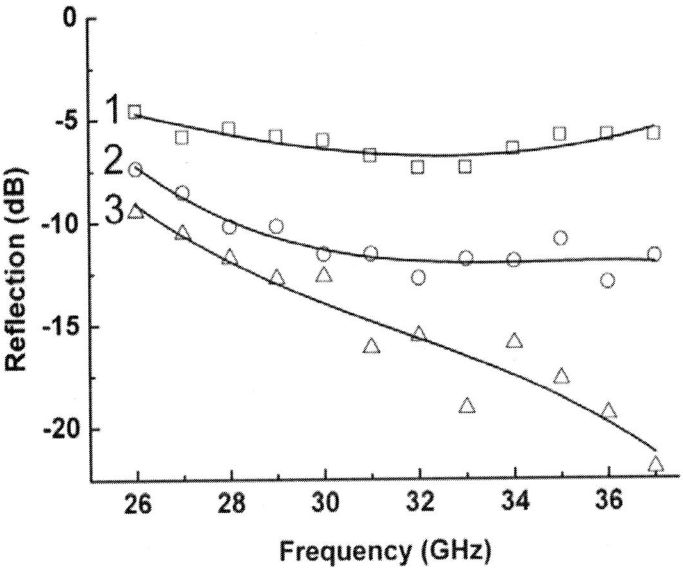

Figure 6. Reflection coefficients from RAC on the metal surface. Curve 1: sample 1 (d = 1 mm). Curve 2: sample 2 (d = 1 mm). Curve 3: two- layer RAC: the sample 1 (d_2 = 1 mm) + sample 2 (d_1 = 1 mm).

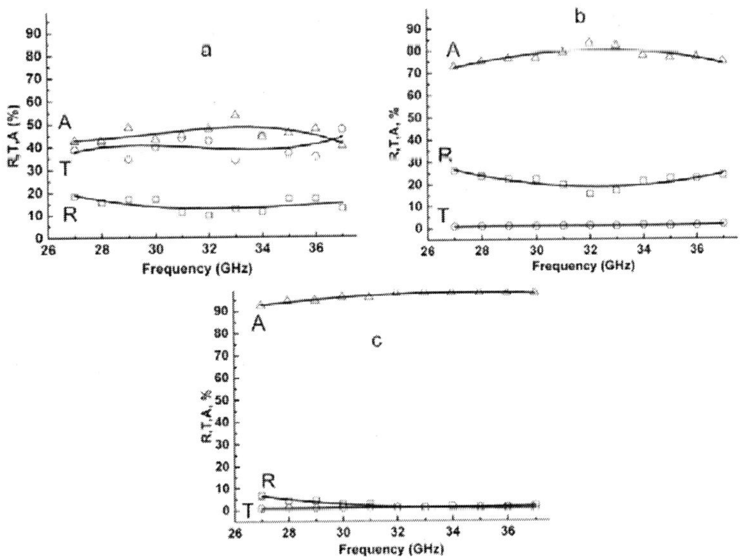

Figure 7. RAC reflection (R), transmission (T), and absorption (A) coefficients. (a) sample 1 (d = 1 mm); (b) sample 2 (d = 1 mm); (c) double layer RAC: the sample 1 (d_2 = 1 mm) + sample 2 (d_1 = 1 mm).

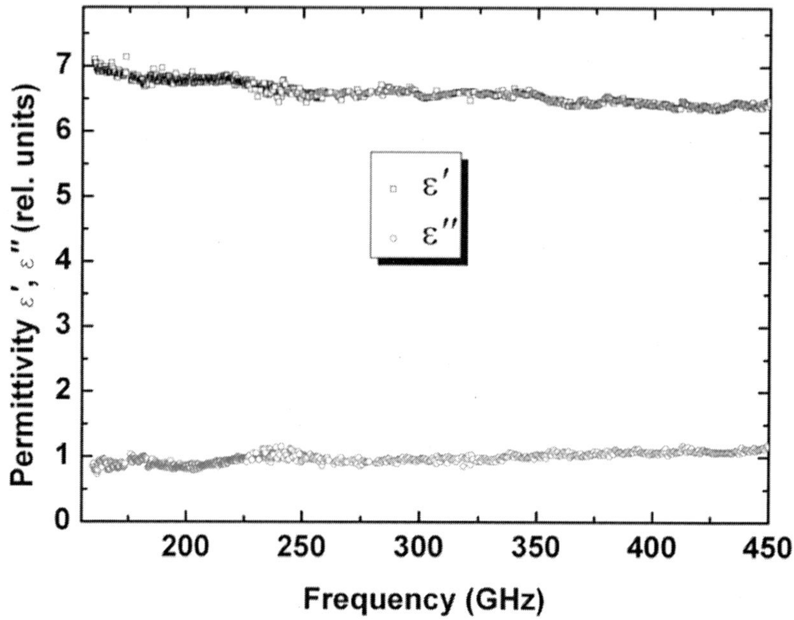

Figure 8. Permittivity spectra of sample 1.

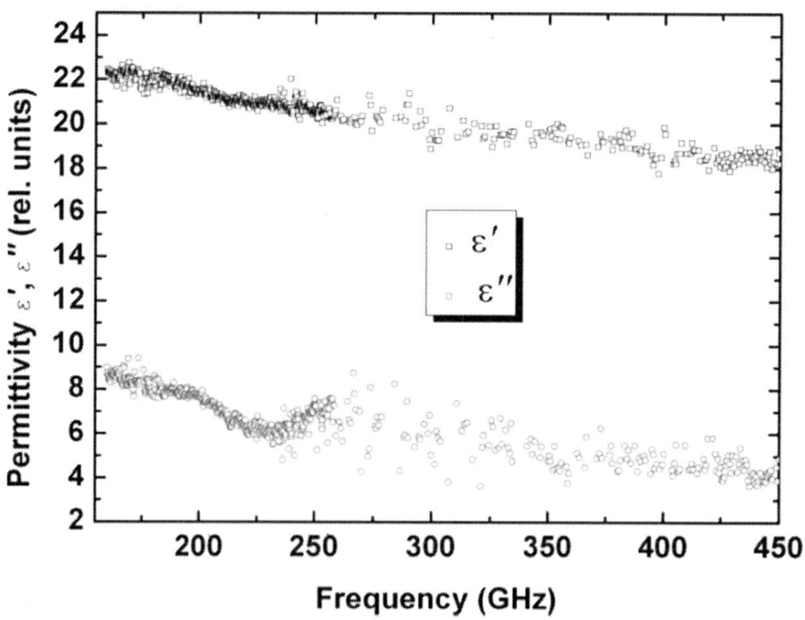

Figure 9. Permittivity spectra of sample 2.

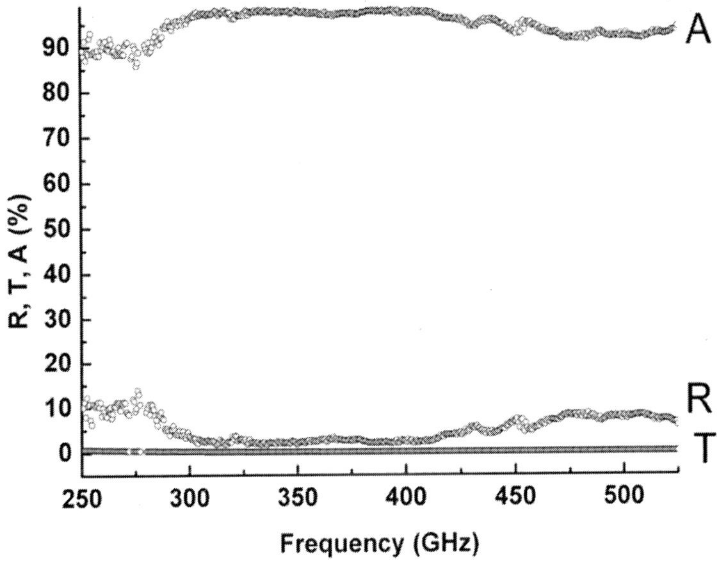

Figure 10. The reflection (R), transmission (T) and absorption (A) coefficients of the sample 1.

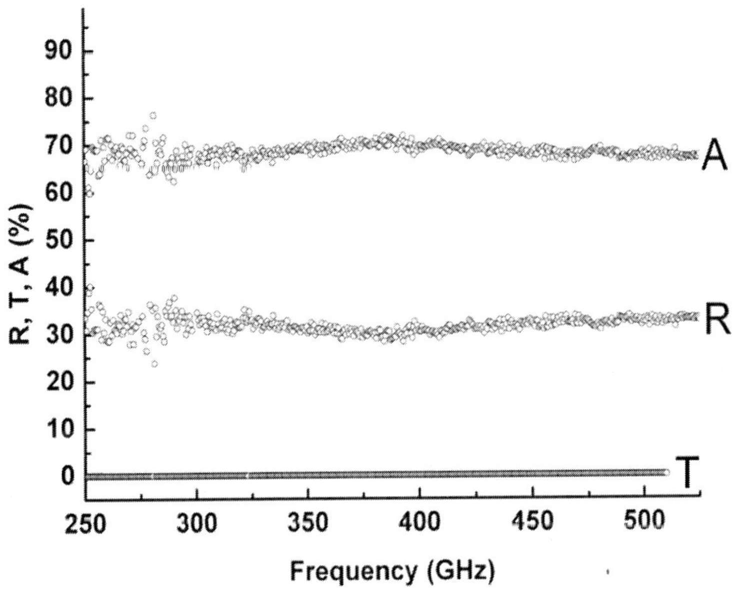

Figure 11. The reflection (R), transmission (T) and absorption (A) coefficients of the sample 2.

4. CONCLUSION

Thus, studies of the permeability and permittivity spectra of the two samples of composites containing CIP in SHF range are carried out. Studies of different designs EMWA have shown that two-layer composite materials based on CIP with different filling factor allow selective absorption with a low reflectance in the SHF and EHF frequency bands. They can be used both as a masking coatings disposed on a metal surface, and as a protective nonreflecting coating with small reflection and transmission coefficients. Sample 1 exhibits good RAC properties also in the frequency range from 250 to 525 GHz.

Acknowledgements

Authors would like to thank center of Tomsk Regional Common Use Center: Center of radio-physics measurements, diagnostic and researching of parameters of natural and artificial materials (e-mail: susl@mail.tsu.ru) [19] .

This work was supported by RFBR grant No. 13-02-12240.

REFERENCES

1. Bystrov, R.P., Dmitriev, V.G., Zemskii, Yu.A., Perunov, Yu.M. and Cherepenin, V.A. (2012) Special Features of the Development of the Radio-Technical Systems of the Radio-Electronic Struggle. Uspekhi Sovremennoi Radioelektroniki (Achievements of Modern Radioelectronics), 8, 3-28.

2. Wang, M., Duan, Y.P., Liu, S.H., Li, X.G. and Ji, Z.J. (2009) Absorption Properties of Carbonyl-Iron/Carbon Black Double-Layer Microwave Absorbers. Journal of Magnetism and Magnetic Materials, 321, 3442-3446.

3. Liu, L., Duan, Y., Liu, S., Chen, L. and Guo, J. (2010) Microwave Absorption Properties of One Thin Sheet Employing Carbonyl-

Iron Powder and Chlorinated Polyethylene. Journal of Magnetism and Magnetic Materials, 322, 1736- 1740.

4. Cheng, Y.L., Dai, J.M., Wu, D.J. and Sun, Y.P. (2010) Electromagnetic and Microwave Absorption Properties of Carbonyl Iron/$La_{0.6}Sr_{0.4}MnO_3$ Composites. Journal of Magnetism and Magnetic Materials, 322, 97-101.

5. Zhuravlev, V.A., Suslyaev V.I. and Korovin, E.Yu. (2010) Dynamic Magnetic Characteristics of a Composite Polymeric Material Based on Carbonyl Iron. Russian Physics Journal, 53, 541-543.

6. Zhuravlev, V.A., Suslyaev V.I., Dotsenko, O.A. and Babinovich, A.N. (2011) Composite Radio-Absorbing Material Based on Carbonyl Iron for Millimeter Wavelength Range. Russian Physics Journal, 53, 874-876.

7. Tan, Y., Tang, J., Deng, A., Wu, Q., Zhang, T. and Li, H. (2013) Magnetic Properties and Microwave Absorption Properties of Chlorosulfonated Polyethylene Matrices Containing Graphite and Carbonyl-Iron Powder. Journal of Magnetism and Magnetic Materials, 326, 41-44.

8. Xu, Y., Zhang, D., Cai, J., Yuan, L. and Zhang, W. (2013) Microwave Absorbing Property of Silicone Rubber Composites with Added Carbonyl Iron Particles and Graphite Platelet. Journal of Magnetism and Magnetic Materials, 327, 82-86.

9. Ding, Q., Zhang, M., Zhang, C. and Qian, T. (2013) Synthesis and Absorbing Mechanism of Two-Layer Microwave Absorbers Containing Polycrystalline Iron Fibers and Carbonyl Iron. Journal of Magnetism and Magnetic Materials, 331, 77-81.

10. Zhang, W., Bie, S., Chen, H., Lu, Y. and Jiang, J. (2014) Electromagnetic and Microwave Absorption Properties of Carbonyl Iron/MnO_2 Composite. Journal of Magnetism and Magnetic Materials, 358-359, 1-4.

11. Gurevich, A.G. and Melkov, G.A. (1996) Magnetization Oscillations and Waves. CRC Press, Boca Raton.

12. Baker-Jarvis, J., Janezic, M.D., Grosvenor Jr., J.H. and Geyer, R.G. (1993) Transmission/Reflection and Short-Circuit Line Methods for Measuring Permittivity and Permeability. NIST Technical Note 1355.

13. Brekhovskikh, L.M. (1973) Waves in Layered Media. Nauka.

14. Belyaev, B.A., Zhuravlev, V.A., Kirichenko, V.I., Suslyaev, V.I. and Tyurnev, V.V. (1992) Use of an Irregular Microstrip Resonator to Investigate Microwave Properties of Dielectrics with Broad Conductivity Ranges. Measurement Techniques, 35, 992-994.

15. Suslyaev, V.I., Kochetkova, T.D., Zhuravlev, V.A. and Sudakov, S.V. (2003) An Automated Setup for Investigating the Temperature Dependence of the Permittivity Spectra of Polar Liquids in the Range of 0.1 - 1.25 GHz. Instruments and Experimental Techniques, 46, 672-676.

16. Naiden, E.P., Suslyaev, V.I., Bir, A.V. and Politov, M.V. (2004) Permeability Spectra Hexaferrite Nanopowders. Journal of Structural Chemistry, 45, 102-105.

17. Zhuravlev, V.A. and Suslyaev, V.I. (2006) Analysis of the Microwave Magnetic Permeability Spectra of Ferrites with Hexagonal Structure. Russian Physics Journal, 49, 1032-1037.

18. Zhuravlev, V.A. and Suslyaev, V.I. (2006) Analysis and Correction of the Magnetic Permeability Spectra of $Ba_3Co_2Fe_{24}O_{41}$ Hexaferrite by Using Cramers-Kronig Relations. Russian Physics Journal, 49, 840-846.

19. The Electronic Resource. http://www.ckp.tsu.ru/ckp_3/

Chapter 7

Features of Electromagnetic Waves Scattering by Surface Fractal Structures

O. Yu. Semchuk[1] and M. Willander[2]

[1] *Chuiko Institute of Surface Chemistry National Academy of Sciences of Ukraine, Ukraine*

[2] *Department of Science and Technology LInkoping University, Sweden*

1. INTRODUCTION

Accurate measurement of surface roughness of machined workpieces is of fundamental importance particularly in the precision engineering and manufacturing industry. This is mainly due to the more stringent demand on material quality as well as the miniaturization of product components in these industries [1-3]. For instance, in the disk drive industry, to maintain the quality of the electrical components mounted on an optical disk, the surface roughness of the disk must be accurately measured and controlled. Hence, the surface finish, normally expressed in terms of surface roughness, is a critical parameter used for the acceptance or rejection of a product.

Surface roughness is usually determined by a mechanical stylus profilometer. However, the stylus technique has certain limitations: the mechanical contact between the stylus and the object can cause deformations or damage on the specimen surface and it is a pointwise measurement method and is time consuming. Hence a noncontact and more speed optical method would be attractive. Different optical noncontact methods for measuring surface roughness have been developed mainly based on reflected light detection, focus error detection, laser scattering, speckle and the interference method [4-10]. Some of these have good resolutions and are being applied in some sectors where mechanical measuring methods previously enjoyed clear predominance. Among these methods, the light scattering method [11] which is a noncontact area-averaging technique, is potentially more speedy for surface inspection than other profiling techniques particularly the traditional stylus technique. Other commercially available products such as the scanning tunneling microscope (STM), the atomic force microscope (AFM) and subwavelength photoresist gratings [12-15], which are pointwise techniques, are used mainly for optically smooth surfaces with roughnesses in the nanometer range.

In this chapter in the frame of the Kirchhoff method (scalar model) the average coefficient of light scattering by surface fractal structures was calculated. A normalized band-limited Weierstrass function is presented for modeling 2D fractal rough surfaces. On the basis of numerical calculation of average scattering coefficient the scattering indicatrises diagrams for various surfaces and falling angles were calculated. The analysis of the diagrams results in the following conclusions: the scattering is symmetrically concerning a plane of fall; with increase the degree of a surface calibration the picture becomes complicated; the greatest intensity of a scattering wave is observed in a mirror direction; there are other direction in which the bursts of intensity are observed.

2. FRACTAL MODEL FOR TWO-DIMENSIONAL ROUGH SURFACES

At theoretical research of processes of electromagnetic waves scattering selfsimilar heterogeneous objects (by rough surfaces) is a necessity to use the mathematical models of dispersive objects. As a basic dispersive object we will choose a rough surface. As is generally known, she is described by the function *z(x,y)*of rejections z of points of M of surface from a supporting plane (x,y) (fig.1) and requires the direct task of relief to the surface.

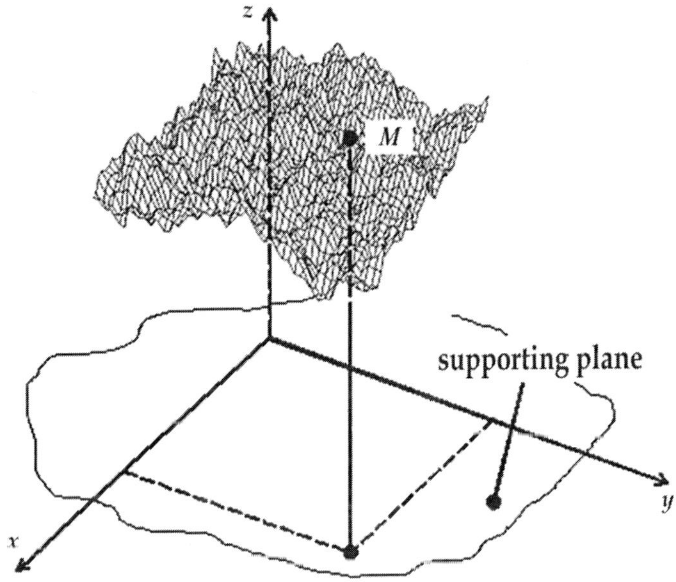

Figure 1. Schematic image of rough surface

There are different modifications of Weierstrass–Mandelbrot function in the modern models of rough surface are used. For a design a rough surface we is used the Weierstrass limited to the stripe function [3,4]

$$z(x,y) = c_w \sum_{n=0}^{N-1} \sum_{m=1}^{M} q^{(D-3)n} \sin\left\{ Kq^n \left[x\cos\frac{2\pi m}{M} + y\sin\frac{2\pi m}{M} \right] + \varphi_{nm} \right\}, \quad (1)$$

where c_w is a constant which ensures that *W(x, y)* has a unit perturbation amplitude; *q*(*q*> 1) is the fundamental spatial frequency; *D* (2 < *D*< 3) is the fractal dimension; *K* is the fundamental wave number; *N* and *M* are number of tones, and $\varphi nm$is a phase term that has a uniform distribution over the interval [$-\pi,\pi$].

The above function is a combination of both deterministic periodic and random structures. This function is anisotropic in the two directions if *M* and *N* are not too large. It has a large derivative and is self similar. It is a multi-scale surface that has same roughness down to some fine scales. Since natural surfaces are generally neither purely random nor purely periodic and often anisotropic, the above proposed function is a good candidate for modeling natural surfaces.

The phases φnm can be chosen determinedly or casually, receiving accordingly determine or stochastic function $z(x,y)$. We further shall consider φnm as casual values, which in regular distributed on a piece $[-\pi;\pi]$. With each particular choice of numerical meanings all $N\times M$ phases φnm (for example, with the help of the generator of random numbers) we receive particular (with the beforehand chosen meanings of parameters cw , q,K,D,N,M) realization of function $z(x,y)$. The every possible realizations of function $z(x,y)$ form ensemble of surfaces.

A deviation of points of a rough surface from a basic plane proportional cw, therefore this parameter is connected to height of inequalities of a structure of a surface. Further it is found to set a rough surface, specifying root-mean-square height of its structure σ , which is determined by such grade:

$$\sigma \equiv \sqrt{\langle h^2 \rangle}, \tag{2}$$

where $h = z(x,y)$, $\langle ... \rangle = \prod_{n=0}^{N-1} \prod_{m=1}^{M} \int_{-\pi}^{\pi} \frac{d\varphi_{nm}}{2\pi} (...)$ - averaging on ensemble of surfaces.

The connection between cw and σcan be established, directly calculating integrals:

$$\sigma = \left[\prod_{n=0}^{N-1} \prod_{m=1}^{M} \int_{-\pi}^{\pi} \frac{d\varphi_{nm}}{2\pi} z^2(x,y) \right]^{\frac{1}{2}} = c_w \left[\frac{M\left(1 - q^{2N(D-3)}\right)}{2\left(1 - q^{2(D-3)}\right)} \right]^{\frac{1}{2}}. \tag{3}$$

So, the rough surface in our model is described by function from six parameters: *cw* (or), *q*,*K*,*D*,*N*,*M*. The influence of different parameters on a kind of a surface can be investigated analytically, and also studying structures of surfaces constructed by results of numerical accounts of Weierstrass function. Analysis of the surface profiles built by us on results of numeral calculations (fig. 2) due to the next conclusions:

- the wave number *K* sets length of a wave of the basic harmonic of a surface;
- the numbers *N* , *M*, *D*and *q*determine a degree of a surface calibration at the expense of imposing on the basic wave from additional harmonics, and *N*and *M* determine the number of harmonics, which are imposed;
- *D* determines amplitude of harmonics;
- *q*both amplitude, and frequency of harmonics. Let's notice that with increase *N*,*M*,*D* and *q*the spatial uniformity of a surface on a large scale is increased also.

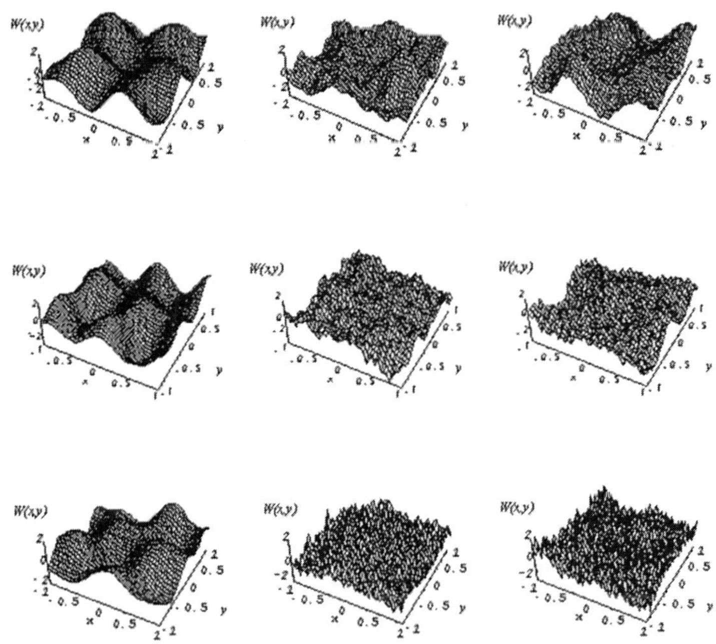

Figure 2. Examples of rough surface by the Weierstrass function K=2π ; N=M=5 ; σ=1. D=2,1; D=2,5; D=2,9(from above to the bottom) q=1,1; q=3; q=7(from left to right)

By means of the original program worked out by us in the environment of Mathematika 5.1 there was the created base of these various types of fractal dispersive surfaces on the basis of Weierstrass function.

Influence each of parameters *q,K,D,N,M* on character of profile of surface it appears difficult enough and determined by values all other parameters. So, for example, at a value *D*=2,1, what near to minimum (*D*=2), the increase of size q does not almost change the type of surface (see the first column on fig.2). With the increase of size D the profile of surface becomes more sensible to the value q (see the second and third columns on fig.2). Will notice that with an increase *N,M,D* and q increases and spatial homogeneity of surface on grand the scale: large-scale "hills" disappear, and finely scale heterogeneities remind a more mesentery on a flat surface.

3. ELECTROMAGNETIC WAVE SCATTERING ON SURFACE FRACTAL STRUCTURES

At falling of electromagnetic wave there is her dispersion on the area of rough surface - the removed wave scattering not only in direction of floppy, and, in general speaking, in different directions. Intensity of the radiation dissipated in that or other direction is determined by both parameters actually surfaces (by a reflectivity, in high, by a form and character of location of inequalities) and parameters of falling wave (frequency, polarization) and parameters of geometry of experiment (corner of falling). The task of this subdivision is establishing a connection between intensity of the light dissipated by a fractal surface in that or other direction, and parameters of surface.

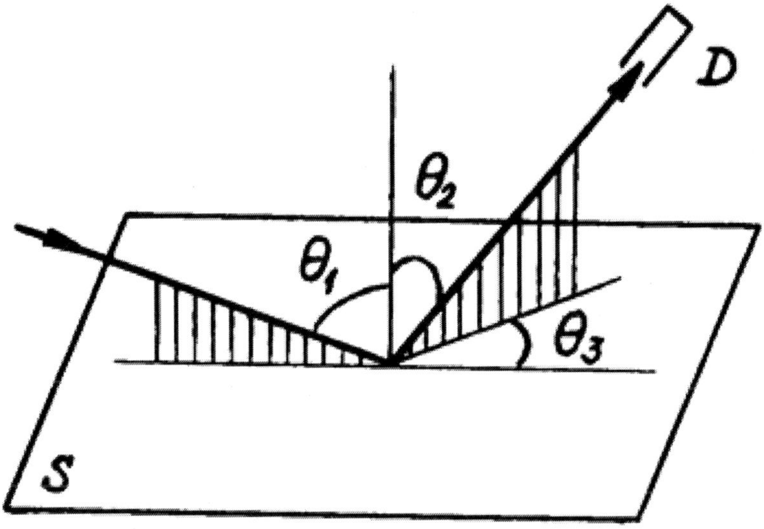

Figure 3. The scheme of experiment on light scattering by fractal surface: S is a scattering surface; D-detector, θ1 is a falling angle; θ2 is a polar angle; θ3 is an azimuthally angle

The initial light wave falls on a rough surface S under a angle $\vartheta 1$ and scattering in all directions. The scattering wave is observed by means of the detector D in a direction which is characterized by a polar angle $\vartheta 2$ and an azimuthally angle $\vartheta 3$. The measured size is intensity of light *Is* scattered at a direction ($\vartheta 2,\vartheta 3$) . Our purpose is construction scattering indicatrise of an electromagnetic wave by a fractal surface (1).

As $Is=\vec{E}s\cdot\vec{E}*s$ (where $\vec{E}s$ is an electric field of the scattering wave in complex representation) that the problem of a finding *Is* is reduced to a finding of the scattered field $\vec{E}s$.

The scattered field we shall find behind Kirchhoff method [16], and considering complexity of a problem, we shall take advantage of more simple scalar variant of the theory according to which the electromagnetic field is described by scalar size. Thus we lose an opportunity to analyze polarizing effects

The base formula of a Kirchhoff method allows to find the scattered field under such conditions:

- the falling wave is monochromatic and plane;
- a scattered surface rough inside of some rectangular ($-X < x_0 < X$, $-Y < y_0 < Y$) and corpulent outside of its borders;
- the size of a rough site much greater for length of a falling wave;
- all points of a surface have the ended gradient;
- the reflection coefficient identical to all points of a surface;
- the scattering field is observed in a wave zone, i.e. is far enough from a scattering surface.

Under these conditions the scattered field is given by

$$E_s(\vec{r}) = -ikrF(\theta_1,\theta_2,\theta_3)\frac{\exp(ikr)}{2\pi r}\int_{S_0}\exp[ik\varphi(x_0,y_0)]dx_0dy_0 + E_e(\vec{r}), \tag{4}$$

Where k is the wave number of falling wave; $F(\theta_1,\theta_2,\theta_3) = -\frac{R}{2C}(A^2+B^2+C^2)$ is a angle factor; R- scattering coefficient; $\varphi(x0,y0)=Ax0+By0+Ch(x0,y0)$ is the phase function; $h(x0,y0)=z(x0,y0)$; $A=\sin\vartheta 1-\sin\vartheta 2\cos\vartheta 3$; $B=-\sin\vartheta 2\sin\vartheta 3$; $C=-\cos\vartheta 1-\cos\vartheta 2$;

$$E_e(\vec{r}) = -\frac{R}{C}\cdot\frac{\exp(ikr)}{4\pi r}(AI_1 + BI_2),$$

$$I_1 = \int_{-Y}^{Y}\left[e^{ik\varphi(X,y_0)} - e^{ik\varphi(-X,y_0)}\right]dy_0,$$

$$I_2 = \int_{-X}^{X}\left[e^{ik\varphi(x_0,Y)} - e^{ik\varphi(x_0,-Y)}\right]dx_0. \tag{5}$$

After calculation of integrals (4) and (5) by means of the formula

$$e^{iz\sin\phi} = \sum_{l=-\infty}^{\infty} I_l(z)e^{il\phi},$$

(where *Il*(*z*)is the Bessel function of the whole order), we receive

$$E_s(\vec{r}) = -2ikFXY\frac{\exp(ikr)}{\pi r}\sum_{l_{\{rs\}}}\left\{\left[\prod_{uv} I_{l_{uv}}(\xi_u)\right]e^{i\sum_{nm} l_{nm}\varphi_{nm}}\right\}\operatorname{sinc}(k_c X)\operatorname{sinc}(k_s Y) + E_e(\vec{r}). \quad (6)$$

where

$$F = F(\theta_1,\theta_2,\theta_3),\ \sum_{l_{\{rs\}}} \equiv \sum_{l_{0,1}=-\infty}^{\infty}\sum_{l_{0,2}=-\infty}^{\infty}\cdots\sum_{l_{(N-1)M}=-\infty}^{\infty},\ \prod_{uv} \equiv \prod_{u=1}^{N-1}\prod_{v=0}^{M},\ \sum_{nm} \equiv \sum_{n=1}^{N-1}\sum_{m=0}^{M},$$

$$\xi_u \equiv kc_w Cq^{(D-3)u},\ \operatorname{sinc} x \equiv \frac{\sin x}{x},\quad \begin{aligned} k_c &\equiv kA + K\sum_{nm} q^n l_{nm}\cos\frac{2\pi m}{M}, \\ k_s &\equiv kB + K\sum_{nm} q^n l_{nm}\sin\frac{2\pi m}{M}, \end{aligned}$$

$$E_e(\vec{r}) = -ikXY\frac{R}{C}\left(A^2+B^2\right)\frac{e^{ikr}}{\pi r}\operatorname{sinc}(kAX)\operatorname{sinc}(kBY).$$

Thus, expression (6) gives the decision of a problem about finding a field scattering by a fractal surface, within the limits of Kirchhoff method.

Now under the formula (4) it is possible to calculate intensity of scattered waves if to set parameters of a disseminating surface *cw* (or) *σD,q,K,N,M,X,Y,ϕnm*, parameter *k* (or *λ=2πk*) a falling wave and parameters *ϑ1,ϑ2,ϑ3*of geometry of experiment. This intensity will be to characterize scattering on concrete realization of a surface *z*(*x,y*) (with a concrete set of casual phases *ϕnm*). For comparison of calculations with experimental data it is necessary to operate with average on ensemble of surfaces intensity $\langle I_s\rangle = \langle \vec{E}_s\vec{E}_s^*\rangle$. Such intensity has appeared proportional intensity $I_0 = \left(\frac{2kXY\cos\Theta_1}{\pi r}\right)^2$ of the wave reflected from the corresponding smooth basic surface, therefore for the theoretical analysis of results it is more convenient to use average scattering coefficient

$$\langle \rho_s \rangle = \frac{\langle I_s \rangle}{I_0}.$$

After calculation $(Is)$and leaving from (6), we shall receive exact expression

$$\langle \rho_s \rangle = \left[\frac{F(\theta_1, \theta_2, \theta_3)}{\cos \Theta_1} \right]^2 \sum_{l\{ns\}} \{ \prod_{nv} I^2_{l_{nv}}(\xi_n) \operatorname{sinc}^2(k_c X) \operatorname{sinc}^2(k_s Y) \} + \\ + \left[\frac{R(A^2 + B^2)}{2C \cos \theta_1} \right]^2 \operatorname{sinc}^2(kAX) \operatorname{sinc}^2(kBY) \tag{7}$$

As expression (7) consist the infinite sum to use it for numerical calculations inconveniently. Essential simplification is reached in case of $\xi_n < 1$. Using thus decomposition function in a line

$$I_v(z) = \left(\frac{3}{2} \right)^v \sum_{k=0}^{\infty} \frac{(-z^2/4)^k}{k!\Gamma(v+k+1)},$$

that rejecting members of orders, greater than ξ_n^2. We shall receive the approached expression for average scattering coefficient

$$\langle \rho_s \rangle \approx \left[\frac{F(\theta_1, \theta_2, \theta_3)}{\cos \theta_1} \right]^2 \{ \left[1 - (k\sigma C)^2 \right] \operatorname{sinc}^2(kAX) \operatorname{sinc}^2(kBY) + \\ + \frac{1}{2} c_f^2 \sum_{nm} q^{2(D-3)n} \operatorname{sinc}^2 \left[\left(kA + Kq^n \cos \frac{2\pi m}{M} \right) X \right] \operatorname{sinc}^2 \left[\left(kB + Kq^n \sin \frac{2\pi m}{M} \right) Y \right] \} + \\ + \left[\frac{R}{2C \cos \theta_1} (A^2 + B^2) \right]^2 \operatorname{sinc}^2(kAX) \operatorname{sinc}^2(kBY), \tag{8}$$

where

$$c_f \equiv k c_w C = k\sigma C \left[\frac{2}{M} \cdot \frac{1 - q^{2(D-3)}}{1 - q^{2N(D-3)}} \right]^{\frac{1}{2}}.$$

4. RESULTS OF NUMERICAL CALCULATIONS

On the basis of numerical calculations of average factor of dispersion under the formula (8) we had been constructed the average scattering coefficient (ρs) from $\vartheta 2$ and $\vartheta 3$.(scattering indicatrix diagrams) for different types of scattering surfaces. At the calculations we have supposed R=1 , and consequently did not consider real dependence of reflection coefficient R from the length of a falling wave λ and a falling angle $\vartheta 1$. The received results are presented on fig. 4.

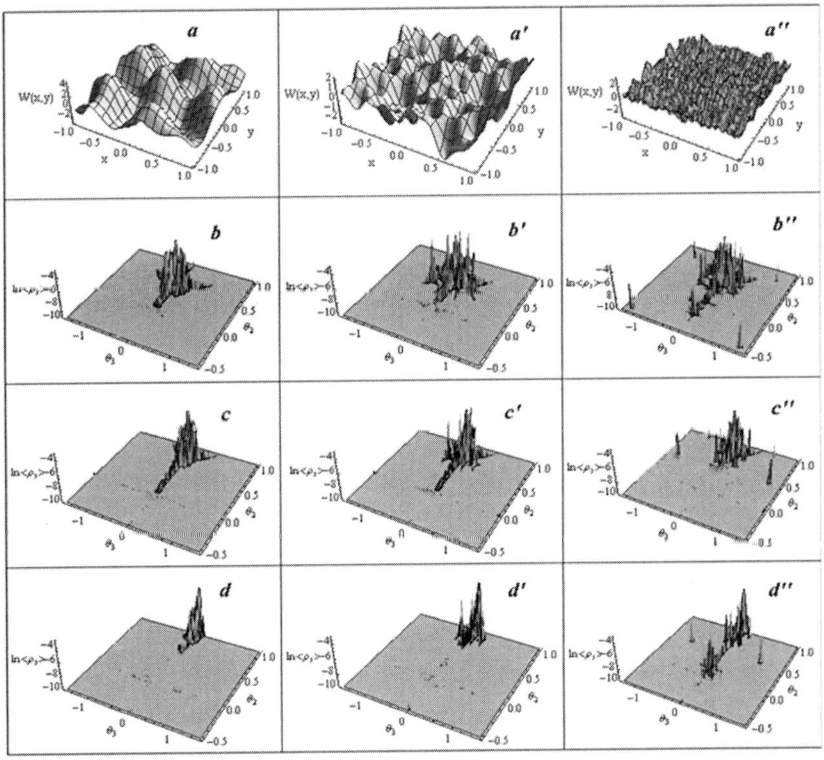

Figure 4. Dependencies of the log(ρs) from the angles θ_2 and θ_3 for the various type of fractal surfaces: a, a', a'' – the samples of rough surfaces, which the calculation of dispersion indexes was produced; from top to bottom the change of scattering index is rotined for three angles of incidence θ1=30,40,600 (a-d, a'-d', a''-d'') at N=5, M=10, D=2.9, q=1.1; n=2, M=3, D=2.5, q=3; N=5, M=10, D.2.5, q=3 accordingly

The analysis of schedules leads to such results:

- Scattering is symmetric concerning of a falling plane;
- The greatest intensity of the scattering wave is observed in a direction of mirror reflection;
- There are other directions in which splashes in intensity are observed;
- With increase in a calibration degree of surfaces (or with growth of its large-scale heterogeneity) the picture of scattering becomes complicated. Independence of the type of scattering surface there is dependence of the scattering coefficient from the incidence angle of light wave. As far as an increase of the incidence angle from 30^0 to 60^0 amounts of additional peaks diminishes. Is their most number observed at $\vartheta_1=30^0$. It is related to influence on the scattering process of the height of heterogeneity of the surface. At the increase of the angle of incidence of the falling light begins as though not to "notice" the height of non heterogeneity and deposit from them diminishes.

The noted features of dispersion are investigation of combination of chaoticness and self-similarity relief of scattering surface.

5. CONCLUSION

In this chapter in the frame of the Kirchhoff method the average coefficient of light scattering by surface fractal structures was calculated. A normalized band-limited Weierstrass function is presented for modeling 2D fractal rough surfaces. On the basis of numerical calculation of average scattering coefficient the scattering indicatrises diagrams for various surfaces and falling angles were calculated. The analysis of the diagrams results in the following conclusions: the scattering is symmetrically concerning a plane of fall; with increase the degree of a surface calibration the picture becomes complicated; the greatest intensity of a scattering wave is observed in a mirror direction; there are other direction in which the bursts of intensity are observed.

REFERENCES

1. T. G. Bifano, H. E. Fawcett, P. A. Bierden, 1997 Precision manufacture of optical disc master stampers. Precis. Eng. 20(1), 53 EOF 62 EOF .
2. P. Wilkinson, et al. 1997 Surface finish parameters as diagnostics of tool wear in face milling, Wear 25(1), 47 EOF 54 EOF .
3. I.'. Sherrington, E. H. Smith, 1986 The significance of surface topography in engineering. Precis. Eng. 8(2). 79 87 .
4. J. Kaneami, T. Hatazawa, 1989 Measurement of surface profiles bv the focus method. Wear 134 221 229 .
5. K. Mitsui, 1986 In-process sensors for surface roughness and their applications. Precis. Eng. 8(40), 212-220.
6. J. W. Baumgart, H. Truckenbrodt, 1998 Scatterometrv of honed surfaces. Opt. Eng. 37(5), 1435-1441.
7. C. J. Tay, L. S. Toh, H. M. Shang, J. B. Zhang, 1995 Whole field determination of surface roughness bv speckle correlation. Appl. Opt. 34(13). 2324 2335 .
8. K. E. Peiponen, T. Tsuboi, 1990 Metal surface roughness and optical reflectance Opt. Laser Technol. 22(2). 127 130 .
9. J. Q. Whitley, R. P. Kusy, M. J. Mayhew, J. E. Buckthat, 1987 Surface roughness of stainless steel and electroformed nickle standards using HeNe laser. Opt. Laser Technol 19(4), 189-196.
10. M. Mitsui, A. Sakai, O. Kizuka, 1988 Development of a high resolution sensor for surface roughness, Opt. Eng. 27(^6). 498 502 .

11. T. V. Vorburger, E. Marx, T. R. Lettieri, 1993 Regimes of surface roughness measurable with light scattering. Appl Opt. 32(19!. 3401 3408 .

12. C. J. Raymond, M. R. Murnane, H. Naqvi, S. S. , J. R. Mcneil, 1995 Metrology of subwavelength photoresist gratings usine optical scat-terometry. J. Vac. Sci. Technol B 13(4), 1484-1495.

13. D. J. Whitehouse, 1991 Nanotechnologv instrumentation. Meas. Control 24(3), 37-46.

14. L. L Madsen, J. F. J.Srgensen, K. Carneiro, H. S. Nielsen, (1993-1994). Roughness of smooth surfaces: STM versus profilometers. Metro-logia 30 513-516.

15. M. Stedman, 1992 The performance and limits of scanning probe microscopes. In Proc. Int. Congr. X-ray Optics and Microanalysis, 347 352 , Manchester. IOP Publishing Ltd.

16. M. V. Berry, Z. V. Levis, 1980 On the Weirstrass-Mandelbrot fractal function. Proc.Royal Soc. London A, 370 459

Chapter 8

Ultrafast Electromagnetic Waves Emitted from Semiconductor

Zhu Yi Ming[1] and Zhuang Song Lin

[1] Engineering Research Center of Optical Instrument and System, Ministry of Education, University of Shanghai for Science and Technology, China

1. INTRODUCTION

Semiconductor devices have become indispensable for generating electromagnetic radiation in every day applications. Visible and infrared diode lasers are at the core of information technology, and at the other end of the spectrum, microwave and radio frequency emitters enable wireless communications. But the ultrafast electromagnetic waves, whose frequency locates in terahertz (THz) region (0.3 – 30 THz; 1 THz = 10^{12} Hz), has remained largely underdeveloped, despite the identification of various possible applications. One of the major applications of THz spectroscopy systems is in material characterization, particularly of lightweight molecules and semiconductors [1] [2]. Furthermore, THz imaging systems may find important niche applications in security screening and manufacturing quality control [3] - [5]. An important goal is the development of three dimensional (3-D) tomographic T-ray imaging systems. THz systems also have broad applicability in a biomedical context, such as the T-ray biosensor [6]. A simple biosensor has been demonstrated for detecting the glycoprotein avidin after binding with vitamin H (biotin) [7].

However, progresses in these areas have been hampered by the lack of efficient ultrafast electromagnetic wave / THz wave sources. As shown in Fig. 1, transistors and other electronic devices based on electron transport are limited to about ~ 300 GHz (~ 50 GHz being the rough practical limit; devices much above that are extremely inefficient) [8]. On the other hand, the wavelength of semiconductor lasers can be extended down to only ~ 10 μm (about ~ 30 THz) [9]. Between two technologies, lie the so called terahertz gap, where no semiconductor technology can efficiently convert electrical power into electromagnetic radiation. The lack of a high power, low cost, portable room temperature THz source is the most significant limitation of modern THz systems. A number of different mechanisms have been exploited to generate THz radiation, such as photocarrier acceleration in photoconducting antennas, second order nonlinear effects in electro-optical (EO) crystals and quantum cascade laser. Currently, conversion efficiencies in all of these sources are very low, and consequently, average THz beam powers tend to be in the nanowatt to microwatt range, whereas the average power of the femtosecond optical source is in the region of ~ 1 W. There is still a long way to go before commercial devices based on this principle can be mass-produced.

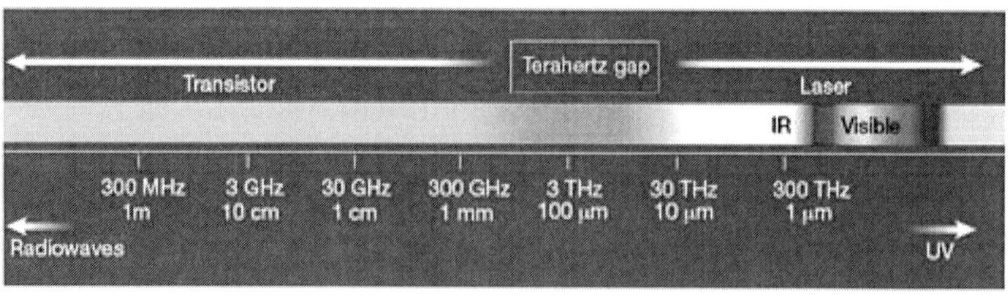

Figure 1. The THz gap. The gap, lying roughly between 300 GHz and 30 THz, exists because the frequencies generated by transistors and lasers, typical semiconductor devices, do not overlap. No current semiconductor technology can efficiently convert electrical power into electromagnetism in this range

Photoconduction is one of the most common approaches for generating broadband pulsed ultrafast electromagnetic beams/THz beams. The photoconductive approach uses high speed photoconductors as transient current sources for radiating antennas [10]. Typical photoconductors include high resistivity GaAs, InP and radiation damaged silicon wafers. Metallic electrodes are used to bias the photoconductive gap and form an antenna. The physical mechanism for THz beam generation in photoconductive antennas begins with an ultrafast laser pulse (with a photon energy larger than the bandgap of the material $h\upsilon \geq E_g$), which creates electron and hole pairs in the photoconductor. The free carriers are then accelerated by the static bias field and form a transient photocurrent, and the fast, time varying current radiates electromagnetic waves. The THz radiation E_{THz}, emitted from the samples should be proportional to the change of the photocurrent *J*:

$$E_{\mathrm{THz}} \propto \frac{\partial J}{\partial t} \tag{1}$$

Figure 2 shows the principle of generating THz radiation from antenna by photoconduction method. Several material parameters affect the intensity and the bandwidth of the resultant THz radiation. For efficient THz radiation, it is desirable to have rapid photocurrent rise and decay time. Thus, semiconductors with small effective electron masses such as InAs and InP are attractive. The maximum drift velocity is also an important material parameter, but it is generally limited by the intraband scattering rate or by intervalley scattering in direct semiconductors such as GaAs [11] [12]. Because the radiating energy mainly comes from stored surface energy in the form of the static bias field, the THz radiation energy scales up with the bias and optical fluency. The breakdown field of the material is another important parameter because this determines the maximum bias that may be applied. Photoconductive emitters are capable of relatively large average THz powers in excess of ~ 40µW [13] and bandwidths as high as ~ 4 THz [14].

As described above, impulsive currents in semiconductors excited by femtosecond optical pulses radiate ultrafast electromagnetic waves in the THz frequency range, serving, for example, as an ultrafast electromagnetic wave source for THz spectroscopy. As described in Eq. (1), the ultrafast electromagnetic wave in the far field is proportional to the time derivative of the current, $\partial J/\partial t$. In a simplified view, since J is proportional to carrier velocity which can be written as $J = nev$, where, e, n, and v are the elementary electric charge, total number of photoexcited carriers and there velocity, respectively, E_{THz} is proportional to carrier acceleration by assuming that the carriers are excited by sufficiently short laser pulses and their lifetime is long enough (n is regarded as being constant). Hence, in principle, the observation of such THz electromagnetic waves provides us with a way of looking into the temporal variation of v, namely the acceleration / deceleration dynamics of photoexcited carriers [11] [12] [15] [16]. Figure 3 shows the illustration of ultrafast electromagnetic waves, THz waves, emitted from intrinsic bulk GaAs photoexcited by femtosecond laser pulses under bias electric fields. By understanding the nonequilibrium transport of carriers in bulk GaAs, the gain region in the THz range, can be clarified, which is the necessary condition for the microwave oscillation [17] [18].

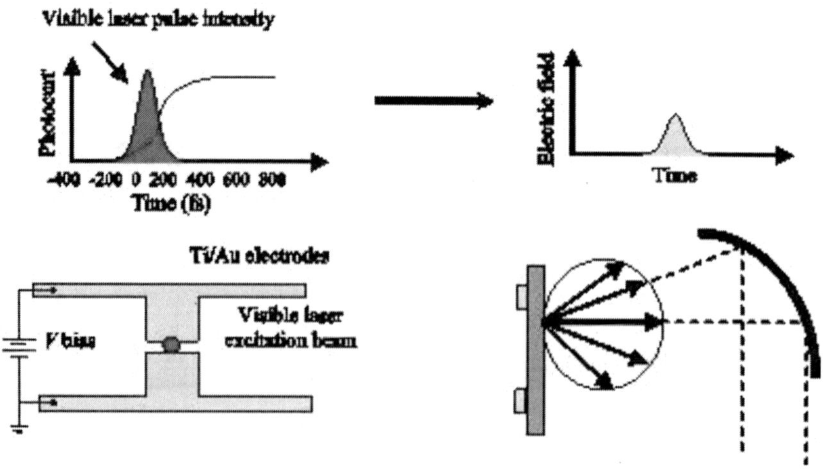

Figure 2. The principle of generating THz radiation from antenna by photoconduction method

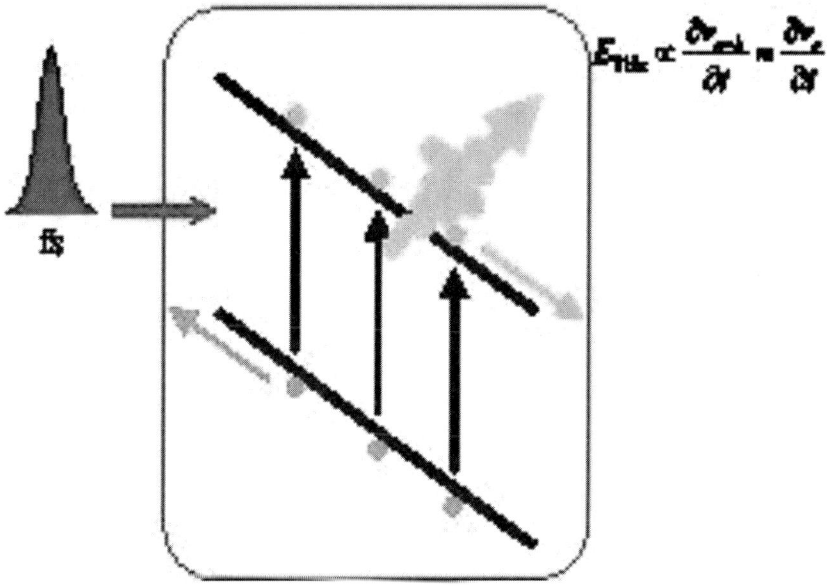

Figure 3. The illustration of generating THz waveforms from intrinsic bulk GaAs photoexcited by femtosecond laser pulses under bias electric fields

The novel experimental technique of the time domain THz spectroscopy for observing ultrafast carrier motion in the femtosecond time regime together with electromagnetic wave radiation that is proportional to dv/dt, is introduced in section 2. Invaluable information, femtosecond acceleration of carriers in bulk GaAs under very high electric fields investigated by time domain THz spectroscopy, contributing to better understanding of device physics in existing high speed electron devices, is discussed in section 3. Finally, the present insights on the gain in GaAs due to electrons intervalley transfer under high electric fields, which is of practical importance for its exploitation in ultrafast electromagnetic wave oscillators, are discussed in section 4.

2. TIME DOMAIN TERAHERTZ SPECTROSCOPY BY ELECTRO-OPTIC SAMPLING METHOD

Time domain terahertz (THz) spectroscopy affords a powerful technique for the research of semiconductors of current industrial interest [2]; the time domain studies of the emitted ultrafast wave / THz wave from various semiconductor structures provide a unique way of looking directly into the temporal as well as spatial evolution of the excited carriers on the sub-picosecond time scale. In this section, time domain THz electro-optic (EO) sampling via the Pockels effect is discussed in detail, including the EO effect, the response spectrum of the EO crystal, and the experimental setup for EO sampling.

Recently, the technique called time domain THz spectroscopy has been attracting wide interest. Time domain THz spectroscopy uses short pulses of broadband THz radiation, which are typically generated using ultrafast laser pulses. This technique has been developed from a work in 1980's at AT&T Bell Labs and IBM T. J. Watson Research Center [19] [20]. Historically, freely propagating THz pulses were measured by means of either photoconductive antennas [21] [22] or far infrared interferometic techniques using incoherent detectors such as bolometers [23] [24]. Although the photoconductive antennas have excellent sensitivity, their frequency response is limited by resonant behavior of the Hertzian dipole structure. For the interferometric techniques, the sensitivity is far worse than that of the photoconductive antennas, because the measurement is coherent and its sensitivity is ultimately limited by the thermal background. In addition, bolometers usually require liquid helium cooling.

On the other hand, ultrafast EO sampling has been widely used in the measurement of local transient electric fields in materials [25] [26]. There existed a need to extend the local electric fields measurement to free space. In 1995, three groups reported their first results using free space EO sampling independently, nearly at the same time [27]-[29]. Although the preliminary results were very poor, rapid progress has been made in the intervening years. It turns out that free space EO sampling is a powerful tool for THz pulse measurement, providing many advantages, such as high sensitivity, ultrabroad frequency response, ease of use, and parallel measurement capability [30]-[35].

Ultrafast carrier motion in the femtosecond time regime accompanies electromagnetic wave radiation, the electric field component of which is proportional to dv/dt. Therefore, the investigation of such electromagnetic wave (or THz radiation) allows us a very unique opportunity of looking directly into the acceleration/deceleration dynamics of carriers in semiconductors. In this chapter, time domain THz spectroscopy based on EO sampling via the Pockels effect is discussed. Furthermore, the response spectrum of the EO crystals are also discussed and 100 μm-thick ZnTe, whose response spectrum is flat up to the frequency around 4 THz, is chosen as the EO crystal for this experiment.

2.1. Electro-optic effect

EO sampling is a phase sensitive detection technique of electromagnetic radiation which measures a birefringence in an EO crystal induced by the incident electromagnetic radiation. This birefringence in the crystal is probed as a phase shift of an optical probe beam. In the time resolved detection of THz pulse, the transient birefringence is probed with the optical pulses at different time delays.

The principle of EO sampling can be explained as follows. Suppose that the probe beam is propagating in the z-direction and *x*, *y* are the crystal axes of the EO crystal as shown in Fig. 4. When an electric field is applied to the EO crystal, the electrically induced birefringence axes *x′* and *y′* are at an angle of 45° with respect to *x* and *y*. If the input beam is polarized along *x*, then the output light can be obtained from the following expression:

$$\begin{pmatrix} E_x \\ E_y \end{pmatrix} = \begin{pmatrix} \cos(\frac{\pi}{4}) & -\sin(\frac{\pi}{4}) \\ \sin(\frac{\pi}{4}) & \cos(\frac{\pi}{4}) \end{pmatrix} \begin{pmatrix} \exp(i\Delta) & 0 \\ 0 & 1 \end{pmatrix} \begin{pmatrix} \cos(\frac{\pi}{4}) & \sin(\frac{\pi}{4}) \\ -\sin(\frac{\pi}{4}) & \cos(\frac{\pi}{4}) \end{pmatrix} \tag{2}$$

where $\Delta = \Gamma_0 + \Gamma$ is the phase difference between the x' and y' polarizations, including both the dynamic (Γ, THz induced) and the static (Γ_0, from the intrinsic or residual birefringence of the EO crystal and the compensator) phase difference. Following Eq. (2), the light intensity in the x and y polarizations is

$$\begin{cases} I_x = |E_x|^2 = I_0 \cos^2 \dfrac{\Gamma_0 + \Gamma}{2} \\ I_y = |E_y|^2 = I_0 \sin^2 \dfrac{\Gamma_0 + \Gamma}{2} \end{cases} \tag{3}$$

where $I_0 = E_0^2$ is the input intensity. It can be seen that I_x and I_y are complementary, i.e. $I_x + I_y = I_0$. This is the result of energy conservation and follows from the fact that absorption in the crystal has been ignored. To extract the light in the x and y polarizations separately, a Wollaston prism is usually used.

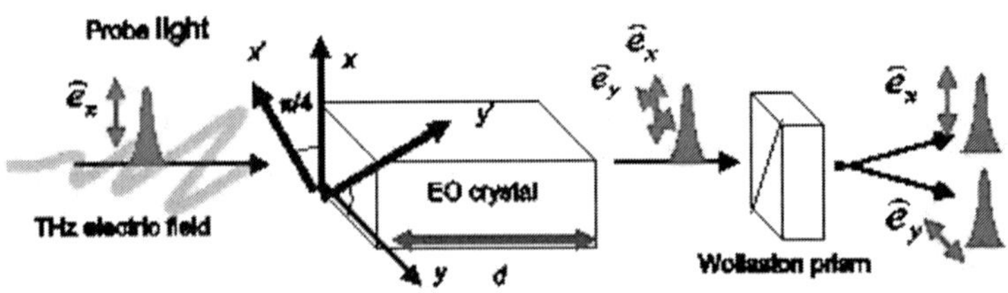

Figure 4. The principle for EO sampling. x' and y' are the birefringence axes of the EO crystal, d is the thickness of the EO crystal

The static phase term Γ_0 (or called the optical bias) is often set equal to $\pi/2$ for balanced detection. For an EO crystal without intrinsic birefringence, such as ZnTe, a quarter wave plate is often used to provide this optical bias. Because $|\Gamma| \ll 1$ in most cases in EO sampling, we have

$$\begin{cases} I_x = \frac{I_0}{2}(1-\Gamma) \\ I_y = \frac{I_0}{2}(1+\Gamma) \end{cases} . \tag{4}$$

In the two beams, the signals have the same magnitudes but opposite signs. For balanced detection, the difference between I_y and I_x is measured, giving the signal

$$I_s = I_y - I_x = I_0\Gamma \tag{5}$$

This signal is proportional to the THz induced phase change, Γ, and, Γ, in turn, is proportional to the electric field of the THz pulse, E_{THz}. For a (110)-oriented ZnTe crystal, the following relation holds:

$$\Gamma = \frac{\pi n_g(\lambda_0)^3 \gamma_{41} d}{\lambda_0} E_{\mathrm{THz}} \tag{6}$$

where d is the thickness of the crystal, $n_g(\lambda_0)$ is the refractive index of the crystal at the wavelength of the optical probe beam λ_0, and γ_{41} is the EO coefficient. By detecting the phase retardation Γ induced in the EO crystal by balanced detection as shown in Eq. (5), E_{THz} can be accurately detected (Eq. (6)).

2.2. The response spectrum of the electro-optic crystal

For a transient THz pulse, phase matching should be considered. When the probe pulse has a group velocity different from that of a THz pulse (so-called GVM), the probe does not always sample the same position on the THz pulse. Instead, it scans across the THz pulse as the two propagating through the crystal, leading to broadening of the measured THz waveform. Furthermore, the bandwidth of EO crystal is limited by reststrahlen band of the crystal. To estimate frequency response and upper limit of the EO sensor, we performed a calculation of the theoretical response function of EO crystals, which shows that the importance of properly accounting for the dispersion of the EO coefficient and the GVM.

The complete frequency response function R(ω) of EO sensors is obtained with the following model: The influences of linear optics such as the GVM between the group velocity of the optical probe pulse and that of the THz pulse, reflection losses at the detector surface, and absorption inside the crystal have been described recently [36] [37].

After a probe pulse and THz wave co-propagate through a sensor of thickness d, the accumulated GVM time is

$$\delta(\omega) = \frac{n_g(\lambda_0) - n(\omega)}{c} d \tag{7}$$

where c is the speed of the light and $n(\omega)$ is the complex refractive index of the THz radiation of frequency ω.

Then, the amplitude and its phase of the EO modulation of the optical probe pulse induced by the THz wave is proportional to d as well as the time average of the electric field across the GVM time, δ (ω). It has the form as [37],

$$G(\omega) = \frac{t(\omega)}{\delta(\omega)} \int_0^{\delta(\omega)} \exp[i2\pi\omega t] dt = t(\omega) \frac{\exp[i2\pi\omega\delta(\omega)] - 1}{i2\pi\omega\delta(\omega)}, \tag{8}$$

where $t(\omega) = 2/[n(\omega)+1]$ is the Fresnel transmission coefficient.

The group refractive index $n_g(\lambda_0)$ at the wavelength of the optical probe pulse is obtained from dispersion data in the near infrared range. The complex refractive index $n(\omega)$ of the THz radiation can be expressed as:

$$n(\omega) = \sqrt{\left[1 + \left\{\frac{(\hbar\omega_{LO})^2 - (\hbar\omega_{TO})^2}{(\hbar\omega_{TO})^2 - (\hbar\omega)^2 - i\hbar\gamma\omega}\right\}\right] \times \varepsilon_\infty} \tag{9}$$

The transverse optical (TO) phonon and longitudinal optical (LO) phonon energies, and the lattice damping are denoted by $\hbar\omega_{TO}$, $\hbar\omega_{LO}$, and γ, respectively. ε_∞ is the high frequency dielectric constant.

$G(\omega)$ describes the detector response if the EO coefficient γ_{41} of the sensor material is frequency independent. However, γ_{41} exhibits a strong dispersion and resonant enhancement due to the lattice resonance. This effect accounts for very pronounced features in the response functions. An analytic expression for $\gamma_{41}(\omega)$ has been derived in [38], Eq. (5):

$$\gamma_{41}(\omega) = r_e \times \left[1 + C\left(1 - \frac{(\hbar\omega)^2 + i\hbar\omega\gamma}{(\hbar\omega_{TO})^2}\right)^{-1} \right] \tag{10}$$

The Faust-Henry coefficient C represents the ratio between the ionic and the electronic part of the EO effect at $\omega = 0$. The purely electronic nonlinearity r_e is assumed to be constant at the mid and far infrared frequencies ω.

The full complex response function $R(\omega)$ of EO sensor is, then, given by

$$R(\omega) = G(\omega) \times \gamma_{41}(\omega)\,. \tag{11}$$

Figure 5 displays the calculated spectra of $|R(\omega)|$ of ZnTe sensors with various thickness, normalized to unity at low frequencies. As shown in Fig. 5, a structured dip in the spectra appears in the reststrahl region around 5.3 THz. This feature reflects the high absorption of the material and the large velocity mismatch near the lattice resonance. At the TO phonon frequency (6.2 THz in ZnTe), the nonlinear coefficient $\gamma_{41}(\omega)$ is resonantly enhanced. We can clearly see, with decreasing the thickness of ZnTe crystal, the upper limit of the frequency increases. However, the sensitivity of the EO sensor is proportional to the thickness of the crystal, as shown in Eq. (6) in section 2.1. Therefore, a trade-off exists between broadband response and high sensitivity.

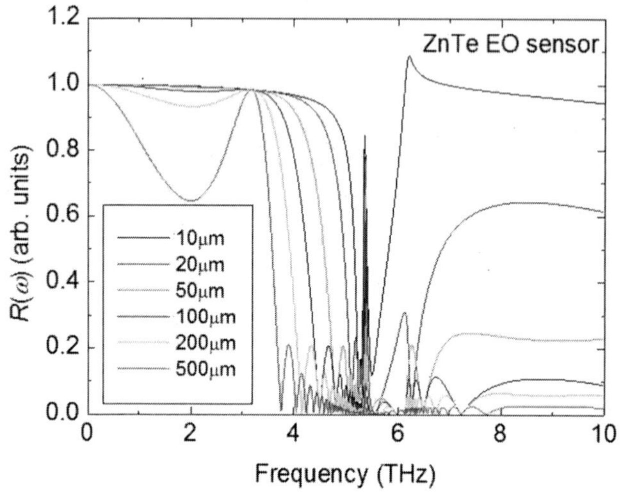

Figure 5. The calculated amplitude response $R(\omega)$ of the ZnTe EO sensor with different thickness, normalized to unity at low frequencies

2.3. Time domain terahertz spectroscopy system

Figure 6 shows the setup for free space THz EO sampling measurements. The ulrafast laser pulse is split by a beam splitter into two beams: a pump beam (strong) and a probe beam (weak). The pump beam illuminates the sample and generates the THz radiation. The generated THz radiation is a short electromagnetic pulse with a duration on the order of one picosecond, so the frequency is in the THz range. The THz beam is focused by a pair of parabolic mirrors onto an EO crystal. The beam transiently modifies the index ellipsoid of the EO crystal via the Pockels effect, as discussed in detail in section 2.1. The linearly polarized probe beam co-propagates inside the crystal with THz beam and its phase is modulated by the refractive index change induced by the THz electric field, E_{THz}. This phase change is converted to an intensity change by a polarization analyzer (Wollaston prism). A pair of balanced detectors is used to suppress the common laser noise. This also doubles the measured signal. A mechanical delay line is used to change the time delay between the THz pulse and the probe pulse. The waveform of E_{THz} can be obtained by scanning this time delay and performing a repetitive sampling measurement. The principle of THz signal sampled by the femtosecond pulse is shown in Fig. 7. By this sampling method, we do not need the instruments for high speed measurement because the THz signals are converted to the electrical signals after the sampling by the femtosecond probe pulse. To increase the sensitivity, the pump beam is modulated by a mechanical chopper and E_{THz} induced modulation of the probe beam extracted by a lock-in amplifier.

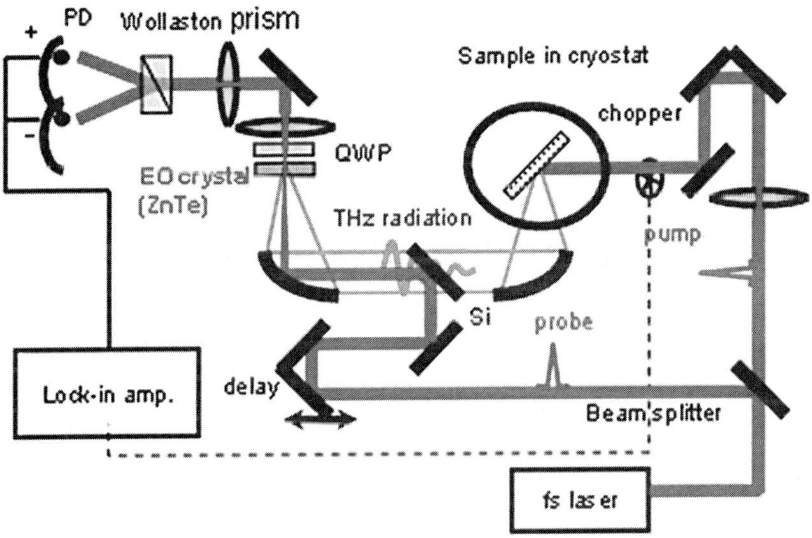

Figure 6. Experimental setup for free space EO sampling

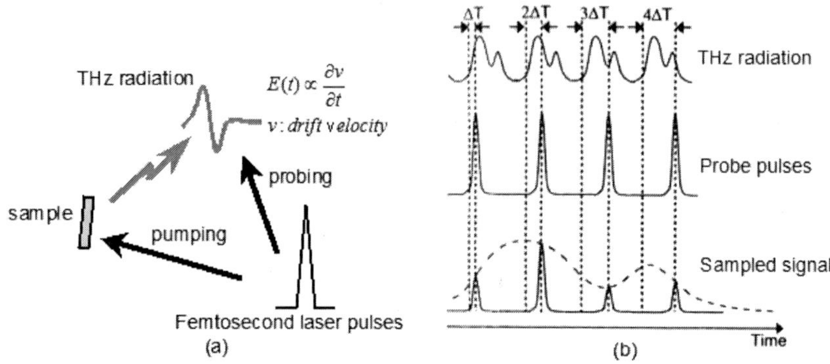

Figure 7. The principle of THz signals sampled by the femtosecond probe beam

3. FEMTOSECOND ACCELERATION OF CARRIERS IN BULK GAAS

Ultrafast nonequilibrium transport of carriers in semiconductors biased at high electric fields is of fundamental interest in semiconductor physics. From a fundamental point of view, the detailed understanding of the femtosecond dynamics of carriers in an electric field is a key issue in many body physics. As an example, one may ask whether the well-developed semiclassical picture of electronic conduction breaks down on ultrafast time scales and at very high electric fields [39]. Furthermore, clarifying carrier dynamics under extremely nonequilibrium conditions is also strongly motivated by the need to obtain information relevant for the design of ultrahigh speed devices. Indeed, transit times even less than 1 ps have been reported for compound semiconductor field effect transistors [40]. In such ultrashort channel transistors, carriers experience very few scattering events in the channel and drift in a very nonstationary manner as schematically illustrated in the Fig. 8. Consequently, the performance of such ultrafast transistors is not mainly determined by the steady-state properties, such as saturation velocities and nobilities, but is governed by the nonstationary carrier transport subjected to high electric fields. It is, therefore, essential to understand nonequilibrium transport of carriers subjected to high electric fields in such devices. However, it has been difficult to characterize such very fast phenomena by using conventional electronics, such as sampling oscilloscopes, because of their limited bandwidth. Consequently, Monte Carlo calculations have been the only tool for discussing transient carrier transport.

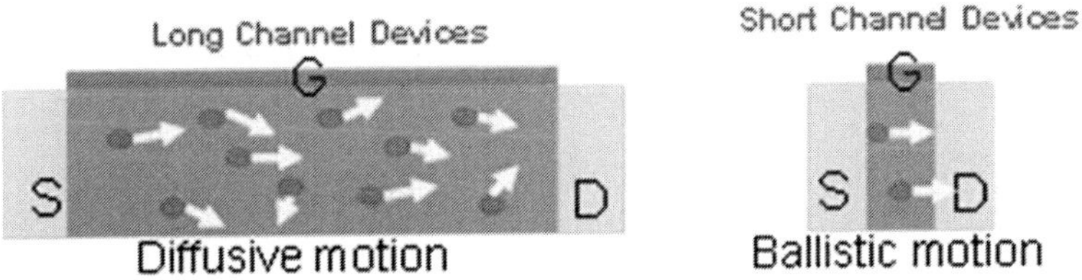

Figure 8. Carrier transport in (a) long diffusive channels (b) very short ballistic channels. (a) In conventional long-channel transistors, their performance is determined by steady-state properties, such as saturation velocities and nobilities. However, (b) In very short channel transistors, carrier transport is quasi-ballistic and their switching speed is governed by how fast carriers are accelerated

So far, only a few experimental studies have been done on the femtosecond carrier acceleration under nonequilibrium in compound semiconductor [11] [12] [15]. Leitenstorfer, et al. presented the first experimental results on the ultrafast electromagnetic wave / terahertz wave emission from electrons and ions in GaAs and InP accelerated by very high electric fields and showed firm experimental evidence of velocity overshoot of electrons in the femtosecond time range [11] [12]. They separated the contributions of electrons and lattices, and determine the transient acceleration and velocity of carriers with a time resolution in the order of 10 fs. The maximum velocity overshoots and traveling distances during the nonequilibrium regime have been determined for the first time.

In this section, we have investigated nonequilibrium acceleration of carriers in bulk GaAs subjected to very high electric fields by time domain THz spectroscopy [16]. It is found that THz emission waveforms have a bipolar feature; i.e., an initial positive peak and a subsequent negative dip. This feature arises from the velocity overshoot. The initial positive peak has been interpreted as electron acceleration in the bottom of Γ the valley in GaAs, where electrons have a light effective mass, while the subsequent negative dip has been attributed to intervalley transfer from the Γ valley to the X and L valleys.

3.1. GaAs *m-i-n* diode & short channel GaAs *m-i-n* diode

Undoped bulk GaAs sample grown by molecular beam epitaxy was used. Sample #1 had an *m-i-n* geometry with a 1 μm-thick intrinsic GaAs layer. An ohmic contact was fabricated by depositing and annealing AuGeNi alloy on the back side of the sample. A semitransparent NiCr Schottky film was deposited on the surface to apply a DC electric field, F, to the intrinsic GaAs layer. The sample was mounted on a copper plate, which was put in the cryostat, for the time domain THz measurements. The illuminated surface area (window area) was approximately 1.5×1 mm^2. The special design of the window size was adopted to avoid the field screening effect.

3.2. Femtosecond acceleration of carriers in GaAs *m-i-n* diode & short channel GaAs *m-i-n* diode

Femtosecond laser pulses from a mode locked Ti: sapphire laser operated at a repetition rate of 76 MHz was used for time domain THz spectroscopy. The full width at half maximum (FWHM) of spectral bandwidth of the femtosecond laser pulses was approximately 20 meV. The central photon energies of the light pulses were set to 1.422 eV at 300 K and 1.515 eV at 10 K, in such a way that electrons were created near the bottom of the conduction band, as well as holes near the top of the valence band. The free space EO sampling technique was used to record temporal waveforms of THz electric fields emitted from the samples [41] [42], as described in chapter 2. The EO sensor used in this experiment was a 100 μm-thick (110)-oriented ZnTe crystal. The spectral bandwidth of this sensor was approximately 4 THz [11] [37], as shown in Fig. 5 in section 2.2. The corresponding resolution is ~ 250 fs.

Figures 9 show temporal waveforms of the THz electric fields emitted from samples #1 (an *m-i-n* diode with a 1 μm-thick intrinsic GaAs layer) [Figs. 9 (a), 9 (b)] for various *F* at 300 K and 10 K, respectively. In [43], the position of $t = 0$ was empirically determined by choosing a position which does not cause artificial jumps in the phase of Fourier spectra of time domain THz data. To determine the position of $t = 0$ more accurately, we adopted a newly developed method, i.e., the maximum entropy method (MEM) [44]-[46]. The position of $t = 0$ set in Figs. 9 has been determined by MEM within an error of ±30 fs, which is limited by the time interval of the recorded points.

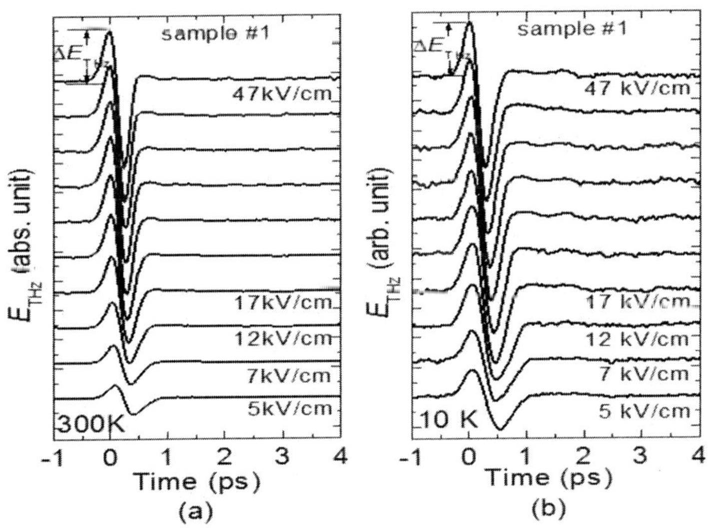

Figure 9. Bias field dependence of the temporal waveforms of THz electric field (E_{THz}) emitted from (a) an *m-i-n* diode with 1 μm-thick intrinsic GaAs layer at 300 K; (b) an *m-i-n* diode with 1 μm-thick intrinsic GaAs layer at 10 K;

From the Maxwell equations,

$$\begin{cases} \nabla \times E = -\mu_0 \dfrac{\partial H}{\partial t} \\ \nabla \times H = J + \dfrac{\partial D}{\partial t} \end{cases} \tag{12}$$

THz electric field, E_{THz}, induced by transient current, J, due to photoexcited carriers is given by

$$E_{\mathrm{THz}} \propto \frac{\partial v}{\partial t} \tag{13}$$

E_{THz} is the emitted ultrafast electromagnetic wave, which is proportional to carrier acceleration. The interband transitions induced by the ultrashort excitation pulses create electrons close to the bottom of the Γ valley in the conduction band as well as holes near the top of the valence band. In an ideal case without scattering, these electrons are ballistically accelerated by the applied electric field in the Γ valley. When electrons in the central valley have gained energy from the applied electric field comparable to the energetic distance between the bottoms of the Γ and the satellite valleys, the electrons are scattered to the satellite valleys by longitudinal optical (LO) phonon. Because the effective mass of electrons in the satellite valley is much heavier than that in the central valley, a sudden decrease of drift velocity is expected. Figure 10 (b) shows the ideal acceleration of electrons in bulk GaAs when the electrons are in the band as shown in Fig. 10 (a). The initial positive part shown in Fig. 10 (b) corresponds to ballistic acceleration of electrons in the Γ valley and the subsequent sudden dip expresses deceleration due to intervalley transfer from the Γ valley to the satellite valleys.

As shown in Fig. 9, the leading edge of the E_{THz} observed in the experiment is due both to the duration of the femtosecond pulses and to the limitation of the spectral bandwidth of the EO crystal detector.

From the experimental data, we can clearly see the THz emission waveforms have a bipolar feature; i.e., an initial positive peak and a subsequent negative dip. This feature arises from the velocity overshoot [12]. The initial positive peak has been interpreted as electron acceleration in the Γ valley, where electrons have a light effective mass, while the subsequent negative dip has been attributed to the electron deceleration due to intervalley transfer from the Γ valley to the X and L valleys. From Figs. 9 (a) and (b), we can find at 10 K, the duration time of the waveforms is longer than that at 300 K at any bias electric fields, *F*, for both samples. The longer duration time of the THz waveforms at 10 K is due to the lower scattering rate of the LO phonon at 10 K, which will be discussed in detail in section 4.2. It is also clearly shown in the Figs. 9 (a) and (b), E_{THz} increases more abruptly with increasing electric field and its bipolar feature becomes pronounced.

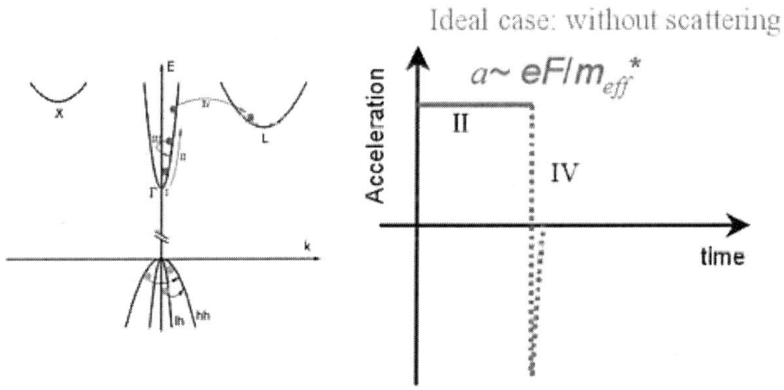

(a) (b)

Figure 10. a) The band diagram of the GaAs; I: Electrons generated at the bottom of the Γ valey; II: Ballistic acceleration; III: Intravalley scattering; IV: Intervalley scattering; (b) ideal acceleration of electrons in bulk GaAs

Taking advantage of the novel experimental method, invaluable information on nonequilibrium carrier transport in the femtosecond time range, which has previously been discussed only by numerical simulations, has been obtained. The present insights on the nonstationary carrier transport contribute to better understanding of device physics in existing high speed electron devices and, furthermore, to new design of novel ultrafast electromagnetic wave oscillators.

4. POWER DISSIPATION SPECTRUM UNDER STEP ELECTRIC FIELD IN BULK GAAS IN TERAHERTZ REGION

It is well known that negative differential conductivities (NDCs) due to intervalley transfer appear in many of the compound semiconductors, such as GaAs, under high electric fields. NDCs are of practical importance, notably for its exploitation in microwave and ultrafast electromagnetic wave / THz wave oscillators [47]. However, since such NDC gain has a finite bandwidth, it gives intrinsic upper frequency limit to ultrafast electromagnetic wave oscillators [48]. For this reason, a number of works have been done to investigate the mechanism, which limits the bandwidth of the gain, and found that it is mainly controlled by the energy relaxation time [49]-[51]. Figure 11 shows the real and imaginary parts of the differential mobility spectra, Re[$\mu(\omega)$] and Im[$\mu(\omega)$], in bulk GaAs for various electric fields at 300 K calculated by using Monte Carlo method [52]. Monte Carlo results predicted that the real part of the negative differential mobility (i.e., gain) can persist up to a few hundred GHz, as seen in Fig. 11. Together with theoretical works, efforts have been devoted to push the high frequency limit into the THz frequency range. Gunn and IMPATT diodes have already been operated at frequencies of 100-200 GHz [53]-[55]. However, in those experiments, high frequency operation was not possible due to electronic circuits; i.e., the cutoff frequency, ν_c, is mainly governed by stray capacitance, inductance, and more fundamentally by the length of the samples, not by the material properties themselves. Therefore, it is extremely important to characterize the gain bandwidth of materials in the THz range.

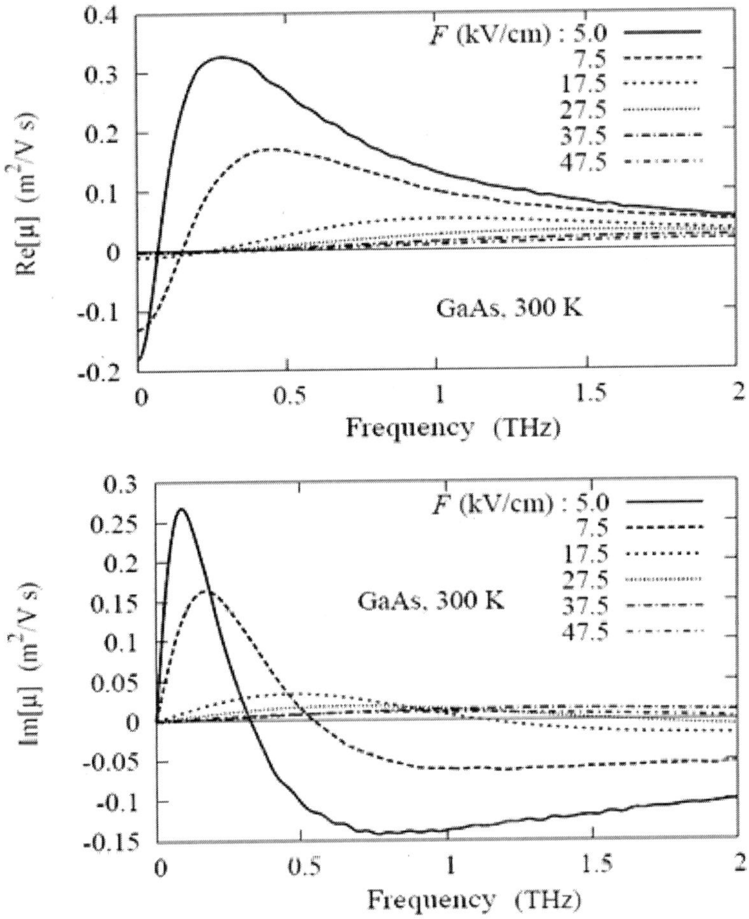

Figure 11. The real and imaginary parts of the differential mobility spectra, Re[μ(ω)] and Im[μ(ω)], in bulk GaAs in THz range for various electric field at 300 K calculated by using Monte Carlo method [52]

The time domain THz spectroscopy has provided us with a unique opportunity to observe the motion of electron wave packet in the sub-picosecond time range and inherently measuring the response of the electron system to the applied bias electric field [11][12][15]. The Fourier spectra of the THz emission give us the power dissipation spectra under step electric field in THz range, from which we can find the gain region of material in THz range.

In this section, we present the measured THz radiation from the intrinsic bulk GaAs under strong biased electric field. The power dissipation spectra under step electric field in THz range have been obtained by using the Fourier transformation of the time domain THz traces [17] [18]. From the power dissipation spectra, the cutoff frequency for negative power dissipation of the bulk GaAs has also been found. It is found that the cutoff frequency for the gain gradually increases with increasing electric fields up to 50 kV/cm and saturates at around 1 THz at 300 K / 750 GHz at 10 K. We also investigated the temperature dependence of cutoff frequency for negative power dissipation, from which we find that this cutoff frequency is governed by the energy relaxation process of electrons from the L valley to the Γ valley via successive optical phonon emission.

4.1. Power dissipation spectrum under step electric field in GaAs

Firstly, we recall the fact that the time domain THz emission experiments inherently measure the step response of the electron system to the applied electric field, as described in more detail in (Shimad et al., 2003). In the THz emission experiment, we first set a DC electric field, *F*, in the samples and, then, shoot a femtosecond laser pulse to the sample at $t = 0$ to create a step-function-like carrier density, nΘ(t), as shown in Fig. 12 (a)-(c). Subsequently, THz radiation is emitted from the accelerated photoexcited electrons [Fig. 12 (d)]. Now, by using our imagination, we view the experiment in a different way. Let's perform the following thought experiment, as shown in Figs. 12 (e)-(h); we first set electrons in the conduction band under a flat band condition [Fig. 12 (f)] and, at $t = 0$, suddenly apply a step-function-like bias electric field, FΘ(t) [Fig. 12 (g)]. We notice that electrons in the thought experiment emit the same THz radiation as in the actual experiment [Fig. 12 (h)]. This fact means that what we do in our femtosecond laser pulse measurement is equivalent to application of a step-function-like electric field to electrons. The step like electric field can be described as,

$$F(t) = F_0\Theta(t) = \begin{cases} 0 & (t<0) \\ F_0 & (t>0) \end{cases} \tag{14}$$

By noting this important implication, the power dissipation spectra under step-function-like electric fields in THz range can be obtained from the thought experimental scheme, as shown in the following;

In time domain, the power density is defined as $P(t) = \frac{\partial \varepsilon}{\partial t} = J(t)F(t)$, i.e., power density equals the product of the current density, *J(t)*, and electric field, *F(t)*. If the direction of *J(t)* is the same as the direction of *F(t)*, this gives a Joule heating. However, if P(t) is negative, this is a gain. Similarly, if the power dissipation spectrum, $S(\omega) = \frac{\partial \varepsilon}{\partial \omega}$, is negative at frequency ω, it means the system gains the energy at this frequency; i.e., gain in the frequency domain. The power dissipation spectrum, *S*(ω), is closely related to the differential conductivity spectrum, as

$$S(\omega) = \sigma(\omega)\left|F(\omega)\right|^2 \tag{15}$$

in the linear response regime. However, this formulation which use small signal conductivity, $\sigma(\omega)$, is not strictly correct in nonlinear response situation. Therefore, we will avoid the formulation using $\sigma(\omega)$ and derive the power dissipation spectrum from the time domain THz data.

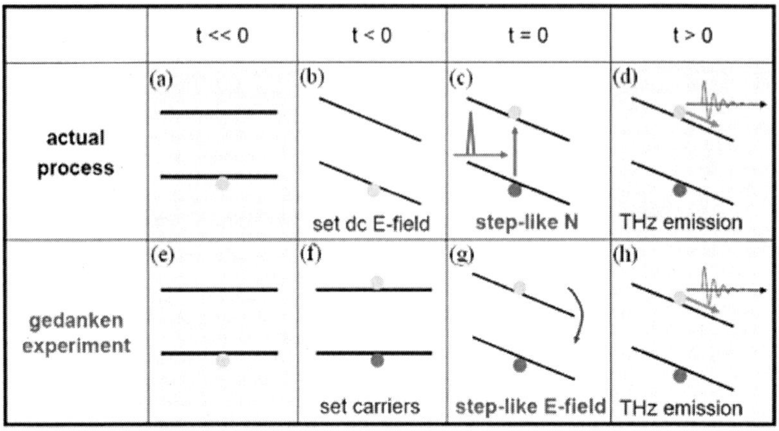

Figure 12. The comparison between the real experiment [(a)-(d)] and thought experiment [(e)-(h)]. Electrons in the thought experiment emit the same THz radiation as in the actual experiment, which means the time domain THz emission experiments inherently measure the step response of the electron system to the applied electric field

Mathematically, the power dissipation, ε, can be expressed as,

$$\varepsilon = \int_{-\infty}^{+\infty} P(t)dt = V\int_{-\infty}^{+\infty} J(t)F(t)dt \,, \tag{16}$$

where V is the volume of the sample, $J(t)$ and $F(t)$ are the time dependent current density and applied electric field, respectively. On the other hand, the energy can also be expressed in the frequency domain as,

$$\varepsilon = V\int_{-\infty}^{+\infty} J(t)F(t)dt = \frac{V}{2\pi}\int_{-\infty}^{+\infty} J(\omega)\bar{F}(\omega)d\omega \,, \tag{17}$$

where $J(\omega)$ and $F(\omega)$ are the Fourier spectra of $J(t)$ and $F(t)$, respectively. Then, the power dissipation spectrum is obtained as,

$$S(\omega) = \frac{\partial \varepsilon}{\partial \omega} = \frac{V}{2\pi} J(\omega)\bar{F}(\omega) \tag{18}$$

From simple mathematics, *J(ω)* can be obtained from the Fourier transformation of the time domain THz traces as,

$$J(\omega) = \int_{-\infty}^{+\infty} J(t)\exp(-i\omega t)dt \propto \int_{-\infty}^{+\infty} [\int_{-\infty}^{t} E_{\mathrm{THz}}(t)dt]\exp(-i\omega t)dt$$
$$= \frac{1}{i\omega}\int_{-\infty}^{+\infty} E_{\mathrm{THz}}(t)\exp(-i\omega t)dt + \pi E_{\mathrm{THz}}(0)\delta(\omega) = \frac{1}{i\omega}E_{\mathrm{THz}}(\omega) + \pi E_{\mathrm{THz}}(0)\delta(\omega) \quad (19)$$

As mentioned in section 4.3, the creation of step-function-like carrier density by femtosecond laser pulses in the actual experiment can be replaced with the application of step-function-like electric field in the thought experiment scheme, then, the *F(ω)* can be expressed as,

$$F(\omega) = \int_{-\infty}^{+\infty} F(t)\exp(-i\omega t)dt = F_0[\frac{1}{i\omega} + \pi\delta(\omega)], \quad (20)$$

where F_0 is the magnitude of the internal electric field applied on the sample.

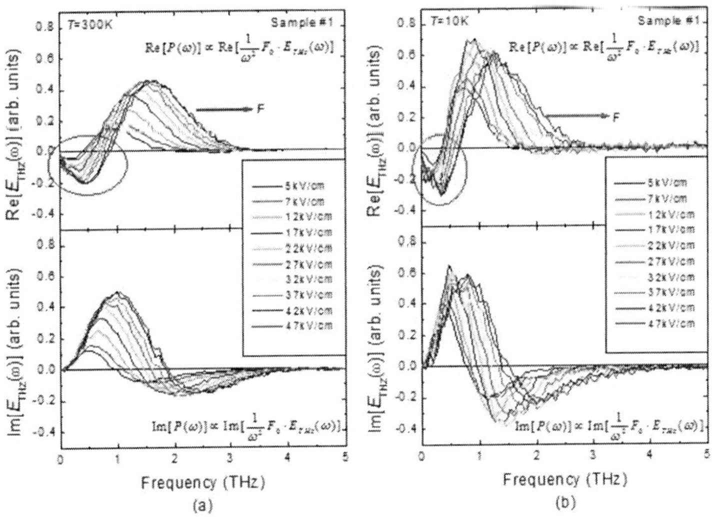

Figure 13. The Re[E_{THz} *(ω)*] and Im[E_{THz} *(ω)*], which are proportional to the real and the imaginary parts of the power dissipation spectra in bulk GaAs for various F at 300K and 10K come from the Fourier transformation of THz traces emitted from sample #1, an *m-i-n* diode sample with a 1 μm-thick intrinsic GaAs layer

Simply substituting Eq. (19) and (20) into Eq. (18), the power dissipation spectrum under step-function-like electric field can be written as,

$$S(\omega) = \frac{F_0}{\omega^2} E_{\mathrm{THz}}(\omega) \quad \text{for } \omega \neq 0 \tag{21}$$

Then, we can obtain an important message from Eq. (21) that the real and imaginary part of the Fourier spectra of $E_{THz}(t)$ (i.e., Re[$E_{THz}(\omega)$] and Im[$E_{THz}(\omega)$]) are proportional to Re[S(ω)] and Im[S(ω)], respectively.

From the discussion above, we know from the measured temporal waveforms of THz electric fields emitted from GaAs samples shown in Figs. 9, the power dissipation spectra in intrinsic bulk GaAs under step-function-like electric fields can be determined by using Fourier transformation of $E_{THz}(t)$. Figures 13 (a) and (b) show Re[$E_{THz}(\omega)$] and Im[$E_{THz}(\omega)$] for sample #1, a metal-intrinsic-n type semiconductor (*m-i-n*) diode sample with a 1 μm-thick intrinsic GaAs layer under various electric fields at 300 K and 10 K, respectively. In the Fourier transformation process, the position of t = 0 is very crucial. Here, we determined t = 0 by using the maximum entropy method [44]-[46]. t = 0 can be determined within an error of ±30 fs, which is limited by the time interval of the recorded points. The phase misplacement can be written as $\Delta\vartheta = \omega\Delta t$. Therefore, the phase error is estimated within an error of < ±π/7, assuming that the frequency of the signal, ω, is mainly around ~ 2.5 THz.

From Re[$E_{THz}(\omega)$] plotted in Fig. 13, we can see the negative region of Re[$S(\omega)$] persists up to several hundred GHz (~ 800 GHz at 47 kV/cm at 300 K), indicating that GaAs has an intrinsic gain region up to the THz range. This frequency is much higher than the gain region predicted by Monte Carlo simulation shown in Fig. 11. The negative power dissipation turns to positive as the frequency increases. Comparing with Re[$E_{THz}(\omega)$] at 10 K and 300 K shown in Figs. 13 (a) and (b), we find magnitude of $S(\omega)$ at 10 K is larger and the width is narrower than those at 300 K at same applied electric fields.

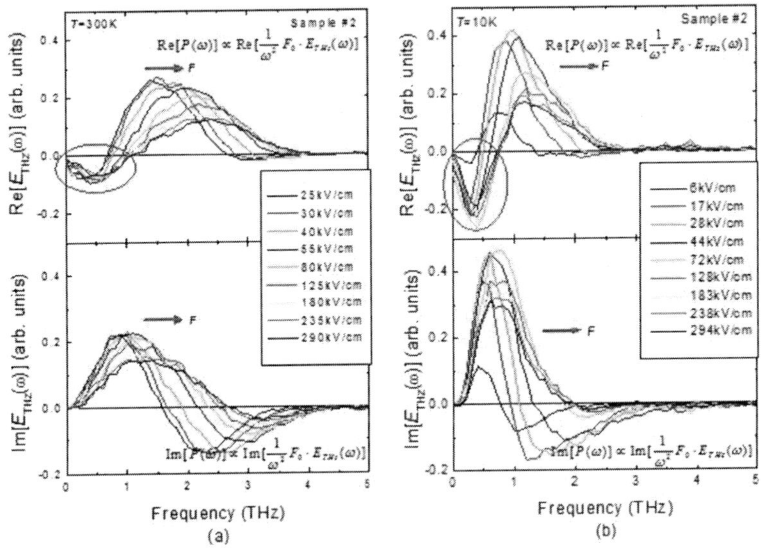

Figure 14. The Re[$E_{THz}(\omega)$] and Im[$E_{THz}(\omega)$], which are proportional to the real and the imaginary parts of the power dissipation spectra in bulk GaAs for various F at 300K and 10K come from the Fourier transformation of THz traces emitted from sample #2, an *m-i-n* diode sample with a 180 nm-thick intrinsic GaAs layer

In the case of collisions with phonons in the Γ valley, each collision redirects the momentum and restrains the drift velocity. Furthermore, intervalley scattering from the Γ valley to the L valley requires phonon with a large wave vector, and group theoretical selection rules show that the longitudinal optical (LO) phonon is the only one allowed because the L valley minimum lies on the edge of the Brillioun zone. Because the probability of LO phonon emission is proportional to <N_{LO}+1> (N_{LO} is the thermal equilibrium phonon population per unit volume; $N_{LO} = \frac{1}{\exp(\hbar\omega_{LO}/k_B T)-1}$) which is temperature dependent, we expect the scattering time is shorter at higher temperatures. The calculation result by Ruch and Fawcett's also showed different magnitude of drift velocity at low and high temperatures [56].

Furthermore, to investigate the power dissipation spectra under extremely large electric fields, we measured the time domain THz traces emitted from sample #2, an *m-i-n* diode with a 180 nm-thick intrinsic GaAs layer, at extremely high electric fields at 300 K and 10 K. Figures 14 (a) and (b) shows the real and imaginary parts of the Fourier spectra of the time domain THz traces emitted from samples #2. We can see even at very high electric fields, for example, F = 300 kV/cm, the negative Re[$E_{THz}(\omega)$] still exists at low frequency region. At high frequency, the Re[$E_{THz}(\omega)$] changes its sign from negative to positive.

4.2. The cutoff frequency for negative power dissipation in GaAs

4.2.1. Electric field dependence of the cutoff frequency

From the real part of the power dissipation spectra, we can find the cutoff frequencies for the gain region, ν_c, at various F. The solid circles and the solid triangles in Fig. 15 show ν_c obtained from sample #1 as a function of F at 300 K and 10 K. The open circles and the open triangles correspond to ν_c obtained from sample #2 at 300 K and 10 K, respectively. The cutoff frequency increases with increasing F for F< 50 kV/cm. However, above 50 kV/cm, ν_c starts to saturate. The sharp kink of ν_c at 50 kV/cm shown in Fig. 15 strongly suggests that this is a very critical electric field. This point will be discussed in section 4.3.

4.2.2. Temperature dependence of the cutoff frequency

To clarify the mechanism of the cutoff frequency, ν_c, we also investigated the temperature dependence of ν_c. Figure 16 shows the temperature dependence of ν_c obtained for sample #1 at 7 kV/cm and 27 kV/cm by solid triangles and circles, respectively. We can clearly see that the temperature dependence of ν_c agrees with that of the emission rate of LO phonons, $<N_{LO}+1>$ (plotted by a dashed line in Fig. 16). The agreement between the experimental results and the LO phonon emission rate strongly suggests that ν_c is governed by the energy relaxation process of electrons from the L valley to the Γ valley via optical phonon emission. At lower temperature, LO phonon emission rate decreases and a longer energy relaxation time is expected, which results in a lower v_c at 10 K than at 300 K, as shown in Fig. 15.

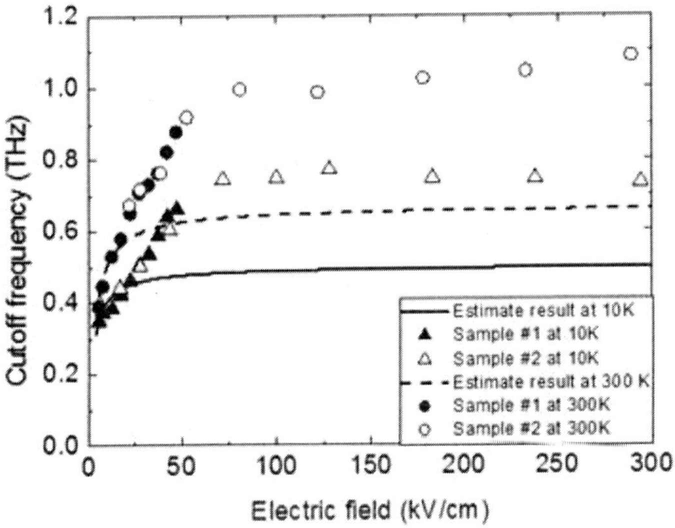

Figure 15. The cutoff frequencies of the negative region of Re[$E_{THz}(\omega)$] obtained for sample #1, as a function of the electric field F at 300 K and 10 K are shown by solid circles and triangles, respectively. The open circles and triangles correspond to the cutoff frequencies obtained from sample #2, at 300 K and 10 K, respectively. The estimated cutoff frequencies are also plotted by a dashed line and a solid line at 300K and 10K, respectively

4.3. Mechanism for the cutoff frequency

When a small step-function electric field is added on a large DC electric field, the electron velocity is predominantly redirected, which causes a general increase of the kinetic energy, or a heating, of the electron system until the mean collision rate with the lattice has risen sufficiently to balance the increasing supply of energy to the electrons from the electric field. The characteristic time for this thermal readjustment is known as the energy relaxation time, which is the limiting process for the upper frequency limit of transferred electron devices [49]. However, the mechanism for the cutoff frequency for gain, ν_c, under extremely nonequilibrium condition has been still argued for more than 30 years.

4.3.1. Energy relaxation and momentum relaxation times govern ν_c.

Monte Carlo simulation has been used as a tool to estimate ν_c [52] [57], as shown in Fig. 11. Gruzinskis [58] and Reggiani, et al. [52] calculated the energy relaxation time τ_ε by Monte Carlo simulation and found that τ_ε is ~ 1.5 ps, as shown in Fig. 17. Also, they discussed that ν_c is determined by both energy relaxation time τ_ε and momentum relaxation time τ_v (τ_v is field dependent and $\tau_v << \tau_\varepsilon$, as shown in Fig. 17) [58]. The calculated ν_c is below 500 GHz, as shown in Fig. 11, which is much lower than ν_c obtained by our experiment.

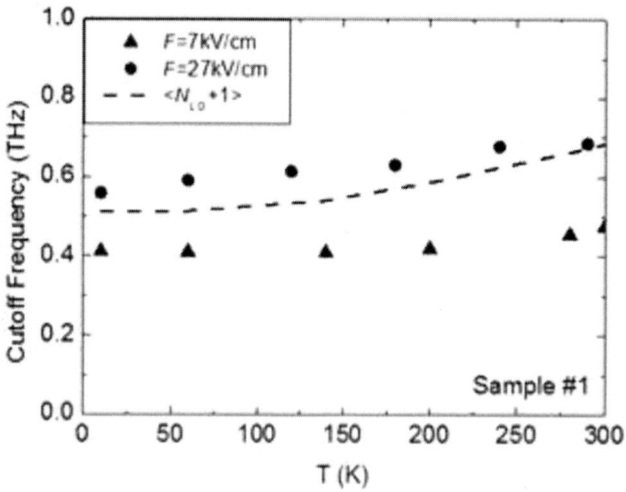

Figure 16. The cutoff frequencies of the negative region of Re[$E_{THz}(\omega)$] obtained for sample #1, as a function of the temperature, at 7 kV/cm and 27 kV/cm are shown by solid triangles and circles, respectively. The reciprocal of the energy relaxation time due to the LO scattering in Γ valley as a function of temperature T is also plotted by a dashed line

4.3.2. Relaxation process of electrons in the L valley governs ν_c.

Another model is that the energy relaxation time is governed by relaxation process of electrons in the L valley [50]. The electrons start to be accelerated at the bottom of the Γ valley, scattered into the L valley, relax and dwell there, and, then, scattered back to the Γ valley. In this model, ν_c is mainly governed by relaxation time of electrons in the L valley and estimated to be ~ 200 GHz, which is also much lower than our experimental result.

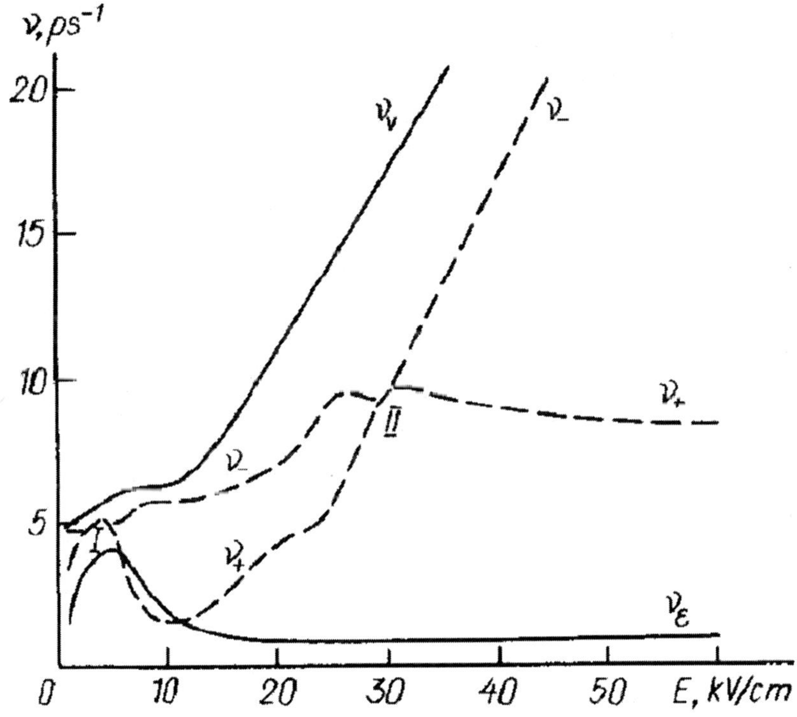

Figure 17. The calculated energy and velocity relaxation rates (solid line) in *n* InP [58]

4.3.3. Relaxation process of electrons in the Γ valley governs ν_c.

The temperature dependence of ν_c, strongly suggests that ν_c is mainly governed by the energy relaxation process of electrons from the L valley to the Γ valley via successive optical phonon emission. Here, we roughly estimate v_c.

In the ε-*k* diagram, the electrons experience a cyclic motion in the *k* space; i.e., be accelerated from the bottom of the Γ valley, go to the L valley by intervalley scattering, be scattered back, and then relax to the bottom of the Γ valley again, as shown in Fig. 18 (a).

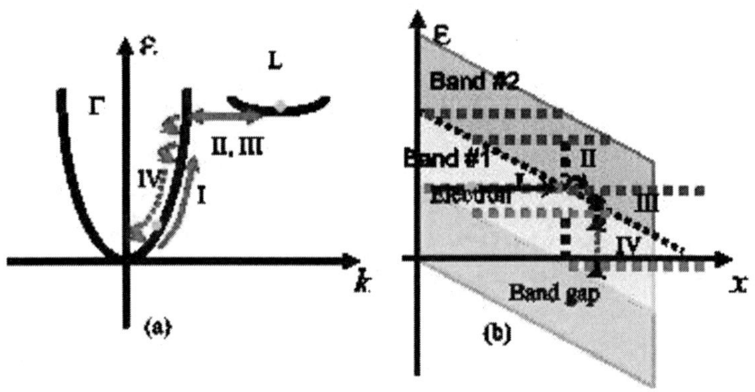

Figure 18. a) Electrons experience a cyclic motion in the *k* space; I: acceleration from the bottom of Γ valley; II: intervalley scattering to the L valley; III: intervalley scattering back to the Γ valley; IV: relaxation to the bottom of the Γ valley; (b) the real space picture; I: acceleration in the band #1 (associated with the Γ valley); II: reach the edge of the band #2 (the L valley related states) and experience intervalley scattering; III: scattering back; IV: relaxation to the bottom of the conduction band

For this order estimation, we assume electrons are ballistically accelerated by the electric field *F* in the Γ valley. In addition, we also neglect the acoustic scattering, *e-e* scattering. Assuming the simple Newton's law

$$\frac{\hbar dk}{dt} = -eF\,, \tag{22}$$

where *e* is the elementary charge, $\hbar = h/2\pi$ (*h* is the Planck constant), *k* is the wave vector of electrons, the time for the acceleration process can be estimated by assuming the energy separation between the Γ minimum and the bottom of L valley $\varepsilon_{\Gamma\text{-}L} = 0.29$ eV. The subsequent intervalley scattering time is about 50 fs [59].

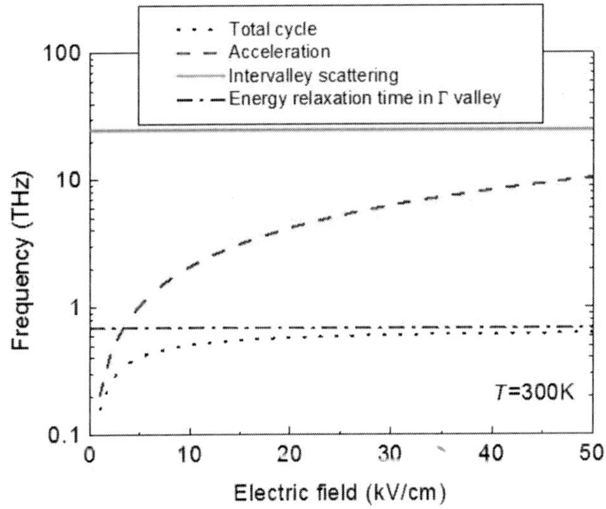

Figure 19. The estimated cutoff frequencies of the negative region of Re[E_{THz} (ω)]as a function of the electric field F at 300 K are shown by black dashed line. The red dashed line, green solid line and blue dash-dotted line correspond to ballistic acceleration process, the intervalley transfer process and the energy relaxation process in Γ valley, at 300 K, respectively

The energy relaxation process in the Γ valley is dominated by the LO phonon scattering. Since the LO phonon energy is 36.5 meV, 8 LO phonons must be successively emitted in the process for electrons to relax from the bottom of the L valley to the bottom of Γ valley ($\varepsilon_{\Gamma\text{-}L}$ = 0.29 eV). The relaxation time for one polar scattering process by LO phonon emission in bulk is expressed [49],

$$\frac{1}{\tau_{(e)}(\bar{k})}=\frac{2\pi}{\hbar}\frac{1}{(2\pi)^3}\frac{e^2\hbar\omega_{LO}}{2\varepsilon_0}\langle N_{LO}+1\rangle\left(\frac{1}{\varepsilon_\infty}-\frac{1}{\varepsilon_r}\right)\frac{m_{eff}{}^*}{\hbar^2\sqrt{\left|\bar{k}\right|^2-\frac{2m_{eff}{}^*\hbar\omega_{LO}}{\hbar^2}}}\times,$$

$$\int_0^{+\infty}\int_0^{\pi}\int_0^{2\pi}\left|k'\right|^2\sin\theta\frac{1}{\left|\bar{k}'-\bar{k}\right|^2}\delta\left(\left|\bar{k}\right|'-\sqrt{\left|\bar{k}\right|^2-\frac{2m_{eff}{}^*\hbar\omega_{LO}}{\hbar^2}}\right)d\left|\bar{k}\right|'d\theta d\phi \tag{23}$$

where ε_0 is the dielectric permittivity of the vacuum, ε_s is the static dielectric constant, ε_∞ is the high frequency dielectric constant, N_{LO} is the thermal equilibrium phonon population in per unit volume, $\hbar = h/2\pi$ (h is the Planck constant), ω_{LO} is the frequency of LO phonon, $m_{eff}{}^*$ is the effective mass of electron, and k, k' are the wave vector of electrons before and after scattering, respectively.

By using Eq. (23), the energy relaxation time in the Γ valley is estimated to be 1.46 ps at 300 K and 1.95 ps at 10 K, by summing up emission times of 8 LO phonons. The temperature dependence of the energy relaxation time is due to the temperature dependence of the phonon emission process, $\langle N_{LO}+1\rangle$. The estimated cutoff frequency, ν_c, at 300 K, are plotted in Fig. 19 [ballistic acceleration (dashed line), intervalley transfer (solid line), energy relaxation (dash-dotted line), and the total (dotted line)]. From this estimation, it is concluded that the energy relaxation process in the Γ valley takes the longest time and governs the upper frequency limitation for the gain. Furthermore, the estimated results are also plotted by a dashed line and a straight line for 300 K and 10 K, respectively, in Fig. 15 for the comparison with the experimental results.

As shown in Fig. 15, the overall trend of this estimation of ν_c is the same as our experimental results. However, at very high electric fields, the magnitude is off by 40%. The discrepancy between experiment and calculation may come from the field dependent energy relaxation time of electrons in the Γ valley as calculated by Fischetti [59]. Fischetti's calculation results show that at very high electric fields, the energy relaxation process is two times faster than that at low electric fields

[59]. Then, the dash-dotted line of Fig. 19 shifts to higher frequency range, which results in higher ν_c.

Under very high electric fields, Wannier Stark (WS) ladder may be formed, although it has never been observed experimentally. The cyclic motion of electrons in the k space can be expressed in the real space as Fig. 18 (b). Electrons are created in the WS ladder states immediately after short pulsed photoexcitation under DC electric field *F*, and accelerated in the conduction band (band #1 in Fig. 18 (b)), which corresponds to the ballistic acceleration in the Γ valley. When they reach the edge of the band #1, they are scattered into the states in the band #2 (L valley related states, as shown in Fig. 18 (b)). After electrons dwell in the upper states, they are scattered downward (intervalley transfer process), and then relax to the bottom of the conduction band (energy relaxation in the Γ valley process).

Based on our experimental data, we conclude that the cutoff frequency is governed by the energy relaxation process of electrons in the Γ valley via successive optical phonon emission. In our estimation, we neglect scattering while electrons are accelerated by the electric field in the band #1. This scattering in the acceleration process makes the sharp kink in the cutoff frequency, ν_c, around 50 kV/cm, as shown in Fig. 15. In the low electric field range ($F < 50$ kV/cm), the number of events of LO phonon scattering decreases with increasing electric field *F*. Each scattering takes ~ 0.13 ps. However, in the high electric field range ($F >$ 50 kV/cm), electrons at the bottom of the conduction band can go into the band #2 without experiencing any scattering. The mean free path of electrons can be estimated by,

$$\lambda = \frac{1}{2}\frac{eF}{m_{eff}^{*}}\tau_v^{2}, \tag{24}$$

where the momentum relaxation time of electrons in band #1 (Γ valley) is about τ_v ~ 0.13 ps, mainly limited by LO phonon scattering. The mean free path is calculated to be ~ 60 nm at F = 50 kV/cm. On the other hand, the distance, *l*, for electrons starting from the bottom of conduction band to the bottom of band #2 is estimated from,

$$eFl = \varepsilon_{\Gamma-L} = 0.29 \text{ eV} . \tag{25}$$

At 50 kV/cm, l is calculated to be ~ 60 nm, almost the same as the mean free path of electrons. Therefore, we can conclude that when *F*< 50 kV/cm, v_c increases with increasing *F* because LO phonon scattering strongly affects the acceleration process, and that for F > 50 kV/cm, v_c is almost *F* independent because no scattering takes place in the acceleration process. The illustration of this situation is shown in Figs. 20.

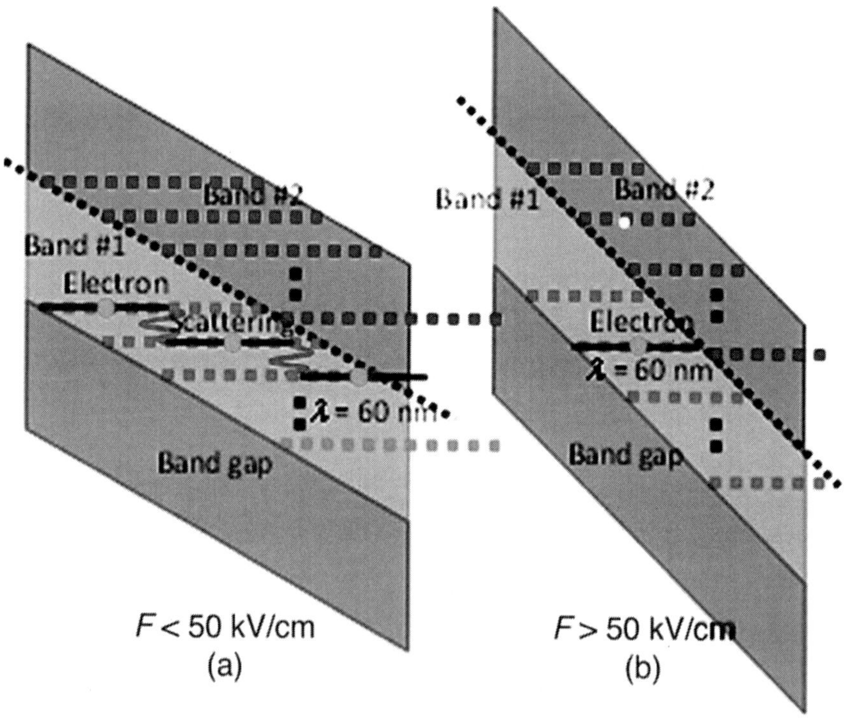

Figure 20. a) For *F*< 50 kV/cm, LO phonon scattering strongly affects the acceleration process; (b) *F*> 50 kV/cm, no scattering takes place in the acceleration process

The other possible explanation is that the strong band mixing gives the sharp kink in ν_c around 50 kV/cm, as shown in Fig. 15. For $F>$ 50 kV/cm, the tunneling probability for electrons in the band #1 to the band #2 is significantly enhanced (Zener tunneling). Such Zener tunneling process may limit the increase in ν_c.

In summary, the power dissipation spectra under step electric field in bulk GaAs in THz range have been investigated by Fourier transformation of the ultrafast electromagnetic waveforms/THz waveforms emitted from intrinsic bulk GaAs photoexcited by femtosecond laser pulses under strong bias electric fields. The cutoff frequency for the gain region, that is of practical importance, notably for its exploitation in ultrafast electromagnetic wave generators, has been also found. The cutoff frequency gradually increases with increasing electric field within 50 kV/cm, saturates and reaches 1 THz (300 K) above 50 kV/cm. Furthermore, illustrated by the electric and temperature dependence of the cutoff frequency, we also find out that this cutoff frequency is mainly dominated by the energy relaxation process of electrons from the L to the Γ valley via successive optical phonon emission.

5. CONCLUSION

Ultrafast electromagnetic waves emitted from semiconductors under high electric fields, which are closely related with ultrafast nonequilibrium transport of carriers in semiconductor, are of fundamental interest in semiconductor physics. Furthermore, clarifying carrier dynamics under extreme nonequilibrium conditions is also strongly motivated by the need to obtain information relevant for the design of ultrahigh speed devices and ultrafast electromagnetic wave emitters. It is, therefore, essential to understand nonequilibrium transport of carriers subjected to high electric fields in such devices. The time domain terahertz (THz) spectroscopy gives us a unique opportunity of observing motions of electron wave packets in the sub-ps time range and measuring the response of electron systems to applied bias electric fields. By understanding the nonequilibrium transport of carriers in bulk GaAs, the intrinsic property of the negative differential conductivity (NDC) in the THz region, which means the gain, can be clarified.

In section 2, we describe the experimental technique of the time domain THz spectroscopy used in this work. Ultrafast carrier motion in the femtosecond time regime accompanies electromagnetic wave radiation that is proportional to dv/dt. Consequently, the investigation of such electromagnetic wave (or THz radiation) allows us a very unique opportunity of looking directly into the acceleration/deceleration dynamics of carriers in semiconductors.

In section 3, the femtosecond acceleration of carriers in bulk GaAs under very high electric fields, F, has been investigated by time domain THz spectroscopy. The observed ultrafast electromagnetic emission waveforms / THz emission waveforms have a bipolar feature; i.e., an initial positive peak and a subsequent negative dip. The initial positive peak has been interpreted as electron acceleration in the bottom of Γ the valley in GaAs, where electrons have a light effective mass, while the subsequent negative dip has been attributed to intervalley transfer from the Γ valley to the X and L valleys.

In section 4, we have investigated the gain due to intervalley transfer under high electric fields, which is of practical importance for its exploitation in ultrafast electromagnetic wave oscillators. The power dissipation spectra under step-function-like electric fields in THz range have been obtained by using the Fourier transformation of the time domain THz traces. From the power dissipation spectra, the cutoff frequencies for negative power dissipation (i.e., gain) in bulk GaAs have been determined. The cutoff frequency gradually increases with increasing electric field, F, for $F<$ 50 kV/cm, saturates and reaches 1 THz (300 K) for $F>$ 50 kV/cm. Furthermore, from the temperature dependence of the cutoff frequency, we found that it is governed by the energy relaxation process of electrons from L to Γ valley via successive optical phonon emission.

Finally, section 5 summarizes this chapter. As demonstrated in this chapter, taking advantage of the novel experimental method, invaluable information on nonequilibrium carrier transport in the femtosecond time range, which has previously been discussed only by numerical simulations, has been obtained. The present insights on the nonstationary carrier transport contribute to better understanding of device physics in existing high speed electron devices. Furthermore, the gain in GaAs due to electrons intervalley transfer under high electric fields obtained from the Fourier transformation of the time domain THz traces is also discussed, which is of practical importance for its exploitation in ultrafast electromagnetic wave oscillators.

6. ACKNOWLEDGEMENTS

The authors thank Mr. JunAn Zhu, a student from the University of Shanghai for Science and Technology, for editing the manuscript of this chapter.

REFERENCES

1. M. van Exter, Fattinger. Ch, D. Grischkowsky, 1989 Terahertz time-domain spectroscopy of water vapour, Opt. Lett. 14 20 1128 1130 .

2. D. Grischkowsky, S. Keiding, M. van Exter, Fattinger. Ch, 1990 Far-infrared time-domain spectroscopy with terahertz beams of dielectrics and semiconductors, J. Opt. Soc. Am. B, 7 10 2006 2015

3. R. A. Cheville, D. Grischkowsky, 1995 Far-infrared terahertz time-domain spectroscopy of flames, Opt. Lett., 20 15 1646 1648

4. B. Ferguson, S. Wang, D. Gray, D. Abbott, X. , C. Zhang, 2002 T-ray computed tomography, Opt. Lett., 27 15 1312 1314

5. B. B. Hu, M. C. Nuss, 1995 Imaging with terahertz waves, Opt. Lett. 20 16 1716 1718

6. S. P. Mickan, A. Menikh, H. Liu, C. A. Mannella, R. Mac, Coll, D. Abbott, J. Munch, X. , C. Zhang, 2002 Label-free bioaffinity detection using terahertz technology, Physics in Medicine and Biology (IOP), 47 21 3789 3795

7. S. P. Mickan, D. Abbott, J.-C Munch, X. Zhang, 2002 Noise reduction in terahertz thin film measurements using a double modulated differential technique, Fluctuation and Noise Letters (Elsevier), 2 1 R13 R28

8. M. J. W. Rodwell, M. Urteaga, T. Mathew, D. Scott, D. Mensa, Q. Lee, J. Guthrie, Y. Betser, S. C. Martin, R. P. Smith, S. Jaganathan, S. Krishnan, S. I. Long, R. Pullela, B. Agarwal, U. Bhattacharya, L. Samoska, M. Dahlstrom, 2001 Submicron scaling of HBTs, IEEE Trans. Electron Devices, 48 11 2606 2624

9. M. Beck, D. Hofstetter, T. Aellen, J. Faist, U. Oesterle, M. Ilegems, E. Gini, H. Melchior, 2002 Continuous Wave Operation of a Mid-Infrared Semiconductor Laser at Room Temperature, Science, 295 5553 301 305

10. G. A. Mourou, C. V. Stancampiano, A. Antonetti, A. Orszag, 1981 Picosecond microwave pulses generated with a subpicosecond laser-driven semiconductor switch, Appl. Phys. Lett. 39 4 295 296

11. A. Leitenstorfer, S. Hunsche, J. Shah, M. C. Nuss, W. H. Knox, 1999 Femtosecond Charge Transport in Polar Semiconductors, Phys. Rev. Lett., 82 25 5140 5142

12. A. Leitenstorfer, S. Hunsche, J. Shah, M. C. Nuss, W. H. Knox, 2000 Femtosecond high-field transport in compound semiconductors, Phys. Rev. B, 61 24 16642 16652

13. G. Zhao, R. N. Schouten, N. van der Valk, W. T. Wenchebach, P. C. M. Planken, 2002 Design and performance of a THz emission

and detection setup based on a semi-insulating GaAs emitter, Rev. Sci. Instrum., 73 4 1715 1719

14. N. Katzenellenbogen, D. Grischkowsky, 1991 Efficient generation of 380 fs pulses of THz radiation by ultrafast laser pulse excitation of a biased metal-semiconductor interface, Appl. Phys. Lett., 58 3 222 224

15. M. Abe, S. Madhavi, Y. Shimada, Y. Otsuka, K. Hirakawa, 2002 Transient carrier velocities in bulk GaAs: Quantitative comparison between terahertz data and ensemble Monte Carlo calculations, Appl. Phys. Lett., 81 4 679 681

16. Y. M. Zhu, T. Unuma, K. Shibata, K. Hirakawa, 2008 Femtosecond acceleration of electrons under high electric fields in bulk GaAs investigated by time-domain terahertz spectroscopy, Appl. Phys. Lett., 93 4 042116 042118

17. Y. M. Zhu, T. Unuma, K. Shibata, K. Hirakawa, 2008 Power dissipation spectra and terahertz intervalley transfer gain in bulk GaAs under high electric fields, Appl. Phys. Lett., 93 23 232102 232104

18. Y. M. Zhu, X. X. Jia, L. Chen, D. W. Zhang, Y. S. Huang, B. Y. He, S. L. Zhuang, K. Hirakawa, 2009 Terahertz power dissipation spectra of electrons in bulk GaAs under high electric fields at low temperature, Acta Phys. Sinica, 58 4 2692 2695

19. D. H. Auston, K. P. Cheung, J. A. Valdmanis, D. A. Kleinman, 1984 Cherenkov Radiation from Femtosecond Optical Pulses in Electro-Optic Media, Phys. Rev. Lett., 53 16 1555 1558

20. Fattinger. Ch, D. Grischkowsky, 1988 Point source terahertz optics, Appl. Phys. Lett., 53 16 1480 1482

21. D. H. Auston, K. P. Cheung, P. R. Smith, 1984 Picosecond photoconducting Hertzian dipoles, Appl. Phys. Lett., 45 3 284 286

22. A. P. De Fonzo, M. Jarwala, C. R. Lutz, 1987 Transient response of planar integrated optoelectronic antennas, Appl. Phys. Lett., 50 17 1155 1157

23. C. Johnson, F. J. Low, A. W. Davidson, 1980 Germanium and germanium-diamond bolometers operated at 4.2 K, 2.0 K, 1.2 K, 0.3 K, and 0.1 K, Opt. Eng., 19 255 258

24. R. C. Jones, 1947 The ultimate sensitivity of radiation detectors, J. Opt. Soc. Am. 37 11 879 890

25. D. R. Dykaar, T. Y. Hsiang, G. A. Mourou, 1985 An application of picosecond electro-optic sampling to superconducting electronics, IEEE Trans. Magn., 21 2 230 233

26. J. A. Valdmanis, G. A. Mourou, 1986 Sub-Picosecond Electro-Optic Sampling: Principles and Applications, IEEE J. Quant. Electron., 22 1 69 78

27. P. U. Jepsen, C. Winnewisser, M. Schall, V. Schya, S. R. Keiding, H. Helm, 1996 Detection of THz pulses by phase retardation in lithium tantalite, Phys. Rev. E 53 4 3052 3054

28. A. Nahata, D. H. Auston, T. F. Heinz, C. Wu, 1996 Coherent detection of freely propagating terahertz radiation by electro-optic sampling, Appl. Phys. Lett., 68 2 150 152

29. Q. X. Wu, C. Zhang, 1995 Free-space electro-optic sampling of terahertz beams, Appl. Phys. Lett. 67 24 3523 3525

30. G. Gallot, D. Grischkowsky, 1999 Electro-optic detection of terahertz radiation, J. Opt. Soc. Am. B, 16 8 1204 1212

31. A. Leitenstrofer, S. Hunsche, J. Shah, M. C. Nuss, W. H. Know, 1999 Detectors and sources for ultrabroadband electro-optic sampling: Experiment and theory, Appl. Phys. Lett., 74 11 1516 1518

32. Z. G. Lu, P. Campbell, X. , C. Zhang, 1997 Free-space electro-optic sampling with a high-repetition-rate regenerative amplified laser, Appl. Phys. Lett., 71 5 593 595

33. C. Winnewisser, P. U. Jepsen, M. Schall, V. Schya, H. Heim, 1997 Electro-optic detection of THz radiation in LiTaO3, LiNbO3 and ZnTe, Appl. Phys. Lett., 70 23 3069 3071

34. Q. Wu, X. , C. Zhang, 1996 Electrooptic sampling of freely propagating terahertz fields, Opt. Quant. Electron., 28 7 945 951

35. Q. Wu, M. Litz, X. , C. Zhang, 1996 Broadband detection capability of ZnTe electro-optic field detectors, Appl. Phys. Lett., 68 21 2924 2926

36. H. J. Bakker, G. C. Cho, H. Kurz, Q. Wu, X. , C. Zhang, 1998 Distortion of terahertz pulses in electro-optic sampling, J. Opt. Soc. Am. B, 15 6 1795 1801

37. Q. Wu, X. , C. Zhang, 1997 7 terahertz broadband GaP electro-optic sensor, Appl. Phys. Lett., 70 14 1784 1786

38. W. L. Faust, C. H. Henry, 1966 Mixing of Visible and Near-Resonance Infrared Light in GaP, Phys. Rev. Lett., 17 25 1265 1268

39. K. K. Thornber, 1978 Path Integrals and Their Applications in Quantum, Statistical and Solid-State physics, Springer, 100306100170 New York

40. Y. Yamashita, A. Endoh, K. Shinohara, K. Hikosaka, T. Matsui, S. Hiyamizu, T. Mimura, 2002 Pseudomorphic In0.52Al0.48As/In0.7Ga0.3As HEMTs with an ultrahigh fT of 562 GHz, IEEE Electron Dev. Lett., 23 10 573 575

41. P. C. M. Planken, H. K. Nienhuys, H. J. Bakker, T. Wenckebach, 2001 Measurement and calculation of the orientation dependence of terahertz pulse detection in ZnTe, J. Opt. Soc. Am. B, 18 3 313 317

42. Q. X. Wu, C. Zhang, 1997 Free-space electro-optics sampling of mid-infrared pulses, Appl. Phys. Lett., 71 10 1285 1286

43. N. Sekine, K. Hirakawa, 2005 Dispersive Terahertz Gain of a Nonclassical Oscillator: Bloch Oscillation in Semiconductor Superlattices, Phys. Rev. Lett., 94 5 057408 067411

44. E. Gornov, E. M. K. Vartiainen, E. Peiponen, 2006 Comparison of subtractive Kramers-Kronig analysis and maximum entropy model in resolving phase from finite spectral range reflectance data, Appl. Opt., 45 25 6519 6524

45. E. M. Vartianen, K. , E. Peiponen, T. Asakura, 1996 Phase Retrieval in Optical Spectroscopy: Resolving Optical Constants from Power Spectra, Appl. Spectrosc., 50 10 1283 1289

46. E. M. Vartiainen, Y. Ino, R. Shimano, M. Kuwata-Gonokami, Y. P. K. Svirko, E. Peiponen, 2004 Nemerical phase correction method for terahertz time-domain reflection spectroscopy, J. Appl. Phys., 96 4171 4175

47. J. B. Gunn, C. A. Hogarth, 1955 A Novel Microwave Attenuator Using Germanium, J.Appl.Physics, 26 353 354

48. P. Das, P. W. Staecker, 1966 Proc. of Sixth Int. Con. on Micr. Gen. & Ampl., Cambridge

49. P. J. Bulman, G. S. Hobson, B. C. Taylor, 1972 Transferred Electron Devices, Academic Press, 100121408507 and New York

50. P. Das, R. Bharat, 1967 Hot electron relaxation times in two-valley semiconductors and their effect on bulk-microwave oscillators, Appl. Phys. Lett., 11 12 386 388

51. A. F. Gibson, T. W. Granville, E. G. S. Paige, 1961 A study of energy-loss processes in germanium at high electric fields using microwave techniques, J. Phys. Chem. Solid 19 3-4 , 198 217

52. L. Reggiani, I. N. F. M. -National, Laboratory. Nano-technology, Ingegneria. Dipartimento di, Innovazione. dell's, Lecce. . Universita di, (Calculated by Professor Reggiani L.)

53. C. Benz, J. Freyer, 1998 200 GHz pulsed GaAs-IMPATT diodes, Electron. Lett., 34 24 2351 2353

54. R. Kamoua, H. Eisele, G. I. Haddad, 1993 D-band (110-170 GHz) InP Gunn devices D-Band (110-170GHz) InP Gunn Devices, Solid-State Electron., 36 11 1547 1555

55. M. F. Zybura, S. H. Jones, B. W. Lim, J. D. Crowley, J. E. Carlstrom, 1996 125-145 GHz stable depletion layer transferred electron oscillators, Solid-State Electron., 39 4 547 553

56. J. G. Ruch, W. Fawcett, 1970 Temperature dependence of transport properties of Gallium Arsenide determined by a Monte-carlo method, J. Appl. Phys., 41 9 3843 3865

57. H. D. Rees, 1969 Hot electron effects at microwave frequencies in GaAs, Soli. Stat. Commun., 7 2 267 270

58. V. Gruzinskis, E. Starikov, P. Shiktorov, 1993 Conservation equations for hot carriers-II. Dynamic features, Solid-State Electron. 36 7 1067 1075

59. M. V. Fischetti, 1991 Monte-carlo simulation of transport in technologically significant semiconductors of the diamond and zincblende structures. 1. Homogeneous transport, IEEE Trans. Electron Devices, 38 3 634 649

Chapter 9

An Overview of the Nineteenth and Twentieth Centuries' Theories of Light, Ether, and Electromagnetic Waves

Salvo D'Agostino

Università "Sapienza", Rome, Italy

ABSTRACT

Johannes Kepler, the seventeenth century celebrated astronomer, considered vision as the effect of its alleged cause—the Lumen. Since many centuries, scientists and philosophers of Light were especially interested in theories and experiments on the cause-effect relationship between our vision and its alleged cause. But the Nineteenth and Twentieth Centuries' contributions of Helmholtz, Maxwell, Hertz, and Lorentz, proved that Light was an electromagnetic wavelike phenomenon, which propagated trough space or ether by an exceptionally high velocity. In my paper I analyze some of the reasons that might justify the controversies among the major experts in Physics and Electrodynamics. In 1905 Albert Einstein found that abolishing Ether would remarkably improve his new Special Relativity theory, and Maxwell's and Hertz's Electrodynamics. His theory was accepted by a large majority of physicists, Max Planck included, but he also found a ten-year silence on the side of Poincaré, and moderate oppositions from Lorentz, the great expert in classical Electrodynamics.

Keywords: Einstein, Maxwell, Helmholtz, Hertz, Lorentz, velocity of Light

1. INTRODUCTION

Johannes Kepler studied what he considered the external cause of vision, the Lumen[1]. Kepler's view of Lumen as a Light's material support represented the birth of a new paradigm on the existence of various levels of Light's supporters. In his 1611 Dioptrics Kepler's theory of vision was based on the concept of rays propagating from luminous and illuminated object. The theory is still accepted as an elementary theory, although Kepler prudently admitted the difficulty of explaining virtual images, and he limited his approach to the study of white light. Kepler's overthrow of the Middle Age conception of Fire as the unique supporter of Light was also confirmed by Galileo's observations with his telescope[2]. Huygens's and Newton's opposite views on the essence of Light concerned a distinction between particles and waves. In the 1820s, Augustine Fresnel was convinced that Light was a transversal wave, and its source was in the Ether, a Lumen very much dissimilar from ordinary materials. An analogy with acoustics and sound was the implicit model because a vibrating string was the causal source of music and sounds. But, the source of Light cannot be assimilated to the string elasticity because the Light's velocity was incommensurably higher than any terrestrial velocity. Maxwell' field theory of electromagnetic waves included an electromagnetic theory of Light, but it owed its success to M.'s recourse to the action at a distance theory of Karl Fredrik Gauss and Wilhelm Weber, clearly opposing his electromagnetic field program. Ten years after Maxwell, Heinrich Hertz performed the first controlled production of electromagnetic waves.

The Discovery of X rays extended the concept of something later on identified as of very high frequency electromagnetic waves. But the spirit of a glorious science did not last for more than a few years: who could have imagined that the very assertor of an indubitable truth in scientific law, the great Max Planck, was destined to contradict Maxwell and Hertz on the wavelike nature of Light? It is well known that Planck discovered that a Light wave included energy packets, the today well known Photons.

In my paper I am concerned with Nineteenth Century's contributions to theories on the velocity of Light by Maxwell, Hertz, Lorentz and Helmholtz. I underline a comparison between Hertz's reflective criticism on his conception of the a-priory assumption of physics theory, and, on the other side, Lorentz's and Zeeman's justifications for the success of their empirically well supported theory. Following the abolition of ether in his celebrated 1905 essay on Special Relativity, Einstein turned many times, and in the majority of his essays, to reconsider the ether's role in physics and concluded his reflections with the admission of the indispensability of ether in his General Relativity Theory.

2. MAXWELL'S THEORY OF LIGHT: A STRUCTURAL SIMILARITY BETWEEN MAXWELL'S ELECTROMAGNETIC WAVES AND THE WAVES OF LIGHT

James Clerk Maxwell brought into his Electromagnetic Theory of Light concepts and experimental data obtained from radically different atomistic and field theories. Theories of Mechanics and Electrostatic Potentials, and Wilhelm's Weber's outstanding contributions to Electrodynamics were largely utilized. Green's and Stokes's elastic and hydrodynamic theories were also the underlying mathematics of M.'s equations for the velocity of electric and magnetic waves. Identification of this velocity with the then known value for the velocity of light, represented the transition from statics to dynamics field theory, a topic which was almost an exclusive merit of the Scottish scientist. It required however a very hard approach, that included elastic, hydrodynamic, and Lagrangian theories.

2. Maxwell's Theory of Light: A Structural Similarity between Maxwell's Electromagnetic Waves and the Waves of Light

Maxwell's contributions to Electromagnetism and Field Theory have been frequently discussed in the history of science literature, with special attention to his original discovery of his celebrated equations[3]. However, to my present knowledge, the connection in Maxwell's research between velocity of his electromagnetic waves and Weber's velocity has not been analyzed, although this is one of the characteristic feature of Maxwell's approach to his theory of Light. The quasi equality between Weber's velocity of motion of electric masses and Maxwell's velocity of electromagnetic waves has been misinterpreted as a logical consequence of premises that were foreign to Maxwell. Maxwell in fact duly underlined Weber's systematic definition of absolute units for electric and magnetic units, although he refused Weber's conception of a velocity of electric masses and only accepted the experimental evidence of Weber's achievements. However, he considered the quasi-equality of Weber's velocity with his electromagnetic waves velocity one of the most important pieces of evidence in favor of his electromagnetic theory of light[4].

The Light velocity in ether was, however, dependent from the choice of Weber's Electrodynamics units, and, in order to meet this problem M. introduced an ad hoc system of three electric and magnetic units. At the end of the Nineteenth Century, the Italian Giovanni Giorgi enlarged the three units system in a more consistent four units system. Maxwell's difficulties with this problem, and his struggle to find arguments in support of his system, are significantly documented in his various attempts to connect Weber's factor to the constants of the Ether, a necessary prerequisite to reach the conclusion that the same factor represented the velocity of electromagnetic waves in Ether[5].

In 1862, he introduced an ad hoc argument from elasticity in order to demonstrate that Weber's factor was the velocity of Light. It is therefore understandable that his later derivation in 1873 from purely metrological arguments of the same demonstration, served to settle the implicit charge of a theoretical circularity[6]. The special form of Maxwell's 1873 demonstration of Weber's equality, based on metrological arguments, afforded a decisive step in the direction of freeing M.'s theory from its early connections with the elastic ether. In fact, let us notice Maxwell's comment in his paper on ether[7]: "....the only ether which has survived is that which was invented by Huygens to explain the propagation of light". And he added: "...the best way to compare the properties of the electromagnetic medium to that of Light is to compare the two velocity... In the form which treats the phenomena of light as the motion of an elastic solid, the wavelike theory is still encumbered with several difficulties"[8].

Although harshly criticized from his contemporaries, M.'s unusual approach relating the velocity of Light to Weber's' factor ratio promoted interesting developments. Between 1900 and 1905, Lord Rayleigh applied with success dimensional analysis to problems of mechanics, heat transfer, optics and electromagnetism. In 1914, E. Buckingham generalized M.'s theorems of products with fixed dimensions in a theory of physically similar systems, the so-called "Π theorem", through which dimensional analysis a useful tool in the theory of physical similitude and of physical models[9].

Writing in 1893, twenty years after M.'s Treatise, Oliver Heaviside grasped the significance of Maxwell's innovations: "Elastic solid theories are a great deal too precise in saying what light consists of, and mechanical speculations in general should be received with much caution, and regarded rather as illustrations or analogies than expressions of facts. We do not know enough yet about the ether for dogmatizing."[10] The essence of Heaviside's statement concerns changes in "permittivity" and "electrical eolotropy". They can be independently observed, i.e., observed through electrical esperiments which do not imply elastic-optical concepts or theories, whereas assertions concerning properties like density and elasticity of ether cannot be observed. The former are observable (factual) properties while the latters remain hypothetical assumptions.

3. A PHENOMENOLOGICAL FOUNDATION FOR AN OBJECTIVE THEORY OF LIGHT IN HELMHOLTZ'S PARALLELISM OF REGULARITIES

Hermann von Helmholtz's (1821-1894) contributions to an epistemological foundation of a theory of vision represent a remarkable part of his work on the Kantian problem of the a-priori in physics[11]. Kant did not excerpt his influence on Helmholtz through the mediation of the romantic philosophy of Shelling and Schopenhauer, less then ever through the Hegelian Naturphilosophie, which Helmholtz vehemently contested. He was inspired by Johannes Müller's theory of the specific energy of "sensory nerves": "The stimulation of the optic nerve produces only sensations of light whether that stimulation be caused by objective light (vibrations in the ether), by an electrical current through the eye, by pressure on the eyeball, or by rapid directional changes of the eye".[12]

By following Mueller's theory, Helmholtz deeply modified the common naive view of perceptions as an image of external influences, such as the Keplerian Lumen. He denied that our senses are passive receptors of a supposedly faithful image of the world: "Inasmuch as the quality of our sensations gives us a report of what is peculiar to the external influence by which it is excited, it may count as a symbol of it, but not as an image. For from an image one requires some kind of alikeness with the object of which it is an image".[13]

His "report" type of perception did not prevent however that Helmholtz found a sort of a likeness of perceptions with the object of science, because in his view "every law of nature" is founded in general on a "parallelism of regularities": "Every law of nature asserts that upon preconditions alike in a certain respect, there always follow consequences which are alike in a certain other respect. Since like things are indicated in our world of sensations by like signs, an equally regular sequence will also correspond in the domain of our sensations to the sequence of like effects by law of nature upon like causes".[14]

Helmholtz believed that his "parallelism of regularities" view of scientific knowledge brought the fullest confirmation of Kant's a-priori doctrine[15]. Let us devote a few lines to illustrate Helmholtz' original view of Kant, because it somehow modified "Kant's doctrine". He added to it a distinction, or resolution, of the forms of intuition into two specific features: a general form and a narrower specification. What was truly a-priori in every sensory perception was a general form devoid of any content, as exemplified in the spatial perception of place, or, in the visual perception as an aggregate of colored surfaces, which always compose our visual field[16]:

"Everything our eye sees, is an aggregate of colored surfaces in the visual field-that is our {general} form of intuition. The particular colours, which appear on this or that occasion -their arrangement and sequence -, this is the result of external influences and is not determined by any law of our makeup."[17].

Let us remark that Helmholtz's study of perceptions as distinguished from a merely linguistic approach, opened a new field of inquiry, that still bears fruit in modern research.

Applying to the study of perceptions in general his distinction of the Kantian a-priori forms in spatial perceptions, Helmholtz found that between the various kinds of sensations there exist differences in the form of two neatly distinguished types: one form concerned sensations belonging to different senses, such as sight and sound, the other to a specific sense as sight; he called this a difference in the modality (of sensations). Inaccessible to Kant's linguistic distinctions, the modality distinction was accessible only via an operational approach[18]. Let us remark that Helmholtz's study of perceptions as distinguished from a merely linguistic approach, opened a new field of inquiry, that still bears fruit in modern research.

On occasion of the anniversary celebration at the Berlin University in 1878, Helmholtz delivered an Address, published by the title "The Facts in Perception"[19]("Die Tatsachen in die Vahrnemung"). He turned his attention to the problem of the correlation between the presumed external agent of Light and the observable effects on our sensation[20]. The problems evidently concerned the old fashioned Keplerian Lumen-Light correlation. In a modern terminology this correlation can be interpreted in a double sense: either in the sense that an ideal theory should achieve a perfect adequacy between concepts and the corresponding perceptions, the Helmholtz' "parallelism", or in the more naturalist view, the existence of ether[21].

Let us look for a possible interpretation at Moritz Schlick's detailed study of Helmholtz's papers. According to Schlick, Helmholtz supported a physiological and psychological interpretation of those forms of intuition that Kant had assigned to a transcendental aspect of knowledge.[22] I argue that by limiting the transcendental character of the Kantian a priori forms of intuition, Helmholtz diminished their cognitive impact, and thus contributed, so-to-speak, to their "secularization." Helmholtz's use of the term "description" referred only to physical phenomena and not to the presumed underlying agents of our perceptions, as the Keplerian Lumen. Notice however that he intended to defend the immediate truth of sensory impressions from attaches by science. Ernst Cassirer rightly considered Helmholtz an important representative of the German phenomenological tradition in physics[23].

4. HEINRICH HERTZ: ETHER POLARIZATION CONFIRMS FARADAY'S VIEW OF AN INDEPENDENT EXISTENCE OF FORCES IN SPACE

One can find agreement and continuity between Helmholtz's and Hertz's theories and conceptions[24], but the historian discovered that their epistemologies present remarkable notes of discordance. In his preface to Hertz's Principles of Mechanics of 1894, in a passage often ignored in the historical literature, Helmholtz inserted a clear statement on the difference between his and Hertz's epistemologies[24]. He remarked that his favorite student had adopted mechanical representations instead of the "simple representation of physical facts and laws by "systems of differential equations," the latter being the phenomenological approach that Helmholtz had embraced[25].

Let us explore those features that could possibly find some ground for his master's charge. Since his initial approach to his experiments, Hertz soon realized that the physical basis of Helmholtz's Electrodynamics did not comport waves, because it was fundamentally an action-at-a-distance theory. In Helmholtz's Electrodynamics in fact, propagation in a medium was considered as a second order effect, which could prevail by a mathematical passage in the so called limiting case, only to comply with some results of Maxwell's theory. In Hertz's criticism, Helmholtz supported a theory of waves as polarization of dielectrics, in the sense of the Poisson-Mossotti polarization theory[26], a peculiar blending of action-by-contact in dielectrics and action-at-a-distance in an ether-less vacuum. Hertz' conception is clearly on the opposite: ether is an-priory concept, indispensable for propagation. Another conceptual obstacle, and a second source of difficulties for Hertz's experiment was his doubt that he did "not always felt quite certain of having grasped the physical significance of his (i.e. Maxwell's) statements". It can be argued that some obscurities were in Maxwell's papers, but the historian can now argue that the main obstacle was Helmholtz: what actually Hertz initially knew on M. theory of Light was mainly derived from Helmholtz' theory.

In my analysis, I find enough evidence for proving that the most relevant topic for Hertz's experimental success was his original conception of ether as a primitive polarized medium[27]. He drastically refuted in fact Helmholtz's thesis that ether was secondarily connected to polarizations of the "Poisson type" (Hertz: "...it is quite another question..."). Hertz suddenly understood that the polarized medium was no less than ether itself, i.e., ether coincided with a polarized dielectrics. If this thesis is accepted, one has an explanation of the otherwise obscure passage in Hertz's Introduction: "I have rather been guided (in my experiments) by Helmholtz's work, as indeed may plainly be seen from the manner in which the experiments are set forth. But unfortunately, in the special limiting case of Helmholtz's theory which leads to Maxwell's equations, and to which the experiments pointed, the physical basis of Helmholtz's theory disappears, as indeed it always does [in Helmholtz] as soon as action-at-a-distance is disregarded. I therefore endeavored to form for myself in a consistent manner the necessary physical conceptions, starting from Maxwell's equations"[28]. The different meaning attributed by Hertz to a polarization theory of ether can be better understood in the light of his conclusions on the nature of electric force, at the end of his work in electromagnetism. His basic tenet was in fact, that air and empty space could support electromagnetic wave not for the reason that they participate in the action of the supposed bound charges- according to a Poisson-Helmholtz conception, but, just on the reverse, air and material dielectrics behave like empty space for the reason that they participate in the nature of empty' space, -i.e. they embed "ether". For the electric force is, essentially, for Hertz a polarization of Ether[29]. One year later, following his experiments, he will attribute to Faraday's glory the new revolutionary concept of a dielectric action, and will express more clearly his ideas on this matter:"The most direct conclusion of the experiment on the finite velocity of propagation of electromagnetic forces, is the confirmation of Faraday's view, according to which the electric forces are polarizations existing independently in space".

A deeper consideration of the process which is at the origin of this Hertz's contribution must take into account, in my opinion, Hertz's first approach to Maxwell's theory in his 1884 theoretical paper[30]: Maxwell's equations are therein deduced from told electrodynamics and from hypotheses that did not imply a dielectric action, in Helmholtz sense. Maxwell's equations result from a combination of Faraday's induction law, energy conservation, and his "principle of the unity of fields", a reiteration process, mathematically equivalent to a series expansion of the field. This fact might have counted in favor of his considering electric waves as dependent on something more essential than the behavior of material dielectrics. An important point in this paper is the so-called "principle of the unity of fields": the electric field has the same nature, irrespective of the mechanism through which it is generated, be it by standing or moving charges or by a static or changing magnetic field. The justification is found in the fact that "... the electric field, according to Faraday's conceptions, is something existing in itself in space independently of the way in which it is generated". It is this physical conception of a field as a self-sustaining entity that makes possible the conception of a self-sustaining wave of electric force as a physical entity (not a purely mathematical one). Hertz's conceptions at this turning point are also manifested by a modification in his outlook upon experiments: he now becomes aware that he is looking for waves of electric force in air. As a consequence, some observations that were disregarded in the past as meaningless became relevant from his new outlook. A further example is given by Hertz in another passage:"Somewhat later on, I thought that I noticed a peculiar reinforcement of the action in front of such shadow forming masses, and of the walls of the room. At first it occurred to me that this reinforcement might arise from a kind of reflection of the electric force from the conducting masses; but although I was familiar with the conceptions of Maxwell's theory, this idea appeared to me to be almost inadmissible so utterly was it at variance with the conceptions then current as to the nature of an electric force". Notice the lack of relevance in observations, when the proper theoretical framework is missing.

As regards Hertz' new conception of electric forces, let us notice that Hertz drastically refuted Helmholtz' atomistic view of electric and magnetic forces. Hertz firmly believed that atoms were in no wise fit for a logically clear and secure foundation of a mathematical theory of physics[31]. In his 1894 Prinzipien, he discussed the problems of atoms in connection with his discussion of Energetics and of its fitness to represent a suitable image (Bild) of physics. He favored Energetics because "it avoids the stumbling-blocks which endangered the appropriateness of the first {image}". i.e., An image (Bild) that represented reality in terms of forces acting between atoms and molecules it is unfit "to serve as a known and secure foundation for mathematical theories". In Hertz's view, "the arbitrary assumed properties of the atoms may not affect the final results, {because} the results may be correct. Nevertheless the details of the deduction are in great part presumably false; the deduction is only in appearance a proof. The earlier modes of thought in physics scarcely allowed any choice or any way of escape".

Overcoming the probable surprise of the modern physicist, let us try to translate Hertz's statements into a modern interpretation: it is impracticable to fit mathematical equations with the types of hypotheses that could represent minute structural properties of matter. Taking account of Hertz's conception of theory and of his views on the mathematics-physics relationship, this means that the form itself of the mathematical structure of Hertz's electrodynamics made impossible its fitting, and on a rational ground, with the atomist hypothesis. In this view, Hertz somehow reflected Maxwell's general ideas on the inconclusiveness of experimental confirmation of "the arbitrary assumed properties of the atoms"[32].

As regards Hertz's views on the nature of his electromagnetic waves, let us notice that his 1884 derivation of M.'s equations, implying the unification of ether with dielectric polarization, and with electricity itself, represented an original initial solution to the problem of the electromagnetic waves physical nature. In fact, Max Planck believed[33] that Hertz's original approach to Maxwell was very worthy of physical and epistemological attention.

5. LORENTZ' ETHER: THE LAST RESORT FOR AN ABSOLUTE VELOCITY OF LIGHT

In 1895, Heindrik Antoon Lorentz published his fundamental work "Versuch einer Theorie der electrischen und optischen Erscheinungen in bewegten Körpern", that included his researches on the optical and electromagnetic effects of dielectrics in motion treated from a unitary point of view. It is worth to remark that he learned his electromagnetism from Helmholtz's hands. For the first time, Maxwell's macroscopic equations were derived in terms of average values of microscopic fields, thus showing that the micro-macro connections were feature of a good theory. In this work, Lorentz called the microscopic charges: "ions", but changed this term to that of "electrons" in 1899, after the 1897 discovery of the particle by J.J. Thomson. From the polarization of the outer lines of Zeeman triplet, he predicted the negative charge of the "Lightion", thus correcting Zeeman's initial conclusion in 1897 of a positive charge. Due to the experimental evidence, Lorentz was perhaps encouraged to ignoring Maxwell's and Hertz's objections, and included his theory of electrons in those molecular and atomic theories that were the targets of the two scientists. To their criticism he opposed the consideration that "the too adventurous physicists, although they may run the risk of being deluded by false prospects, are very often repaid by the boldness of molecular hypotheses"[34]. Although he claimed that his electron theory was an offspring from the great theories of electricity of Faraday and Maxwell, he also underlined the differences. "You all know this theory of Maxwell, which we may call the general theory of the electromagnetic field, and in which we constantly have in view the state of the matter of the medium by which the field is occupied. While speaking of this state, I must call your attention to the curious fact that, although we never loose sight of it, we need by no means go far in attempting an image of it and, in fact, we cannot say much about it. It is true that we may represent to ourselves internal stresses existing in the medium surrounding an electrified body or a magnet that we may think of electricity as of some substance or fluid, free to move in a conductor. Many physicists have done all this and Maxwell himself has set the example. Yet, it must not be considered as really necessary; we can develop the theory to a large extent and elucidate a great number of phenomena, without entering upon speculations of this kind"[34].

Lorentz's objected that "the theory of electrons is to be regarded as an extension to the domain of electricity of the molecular and atomic theories that have proved of so much use in many branches of physics and of chemistry. Like these, it is apt to be viewed un favorably by some physicists who prefer to push their way into new and un explored regions by following these great highways of science which we possess in the laws of thermodynamics".

He justified his method on the ground of a new role of mathematics in physics, and expresses his belief a few lines afterwards: "the mathematical relations have thus come to take a very pre-eminent place".

As often in physics, the ground of his ideas found support in another tradition, that of Hermann Helmholtz and Wilhelm Weber's electrodynamics, a paradigm alternative to that of Maxwell and Hertz. In his 1878 essay: "Concerning the Relations between the Velocity of Propagation of Light and the Density and Composition of Media", he supported Hertz's view of ether, whereas the Poisson type dielectric holds only in ponderable bodies. His radical thesis on the null variation of the Light's velocity relative to the earth's motion is presented in his 1887 essay: "De influence du movement de la terre sur les phenomènes lumineux". The negative result of the Michelson-Morley experiment is justified by the theory of a compensatory effect of a Fitzgerald contraction and an account of an apparent time variation. In an 1891 essay "Electricitet en Ether", taking account of Hertz's experiment, he attempted a conciliation between Helmholtz's and Maxwell's electromagnetism.

His new theoretical approach is presented in his great 1892 Memoire "La theorie électromagnetique de Maxwell et son application aux corps mouvantes". A remarkable balance between a physical electrostatic contraction and the Lorentz transformation of Maxwell's equations explains the Michelson Morley null result. The composite aspect of Lorentz's theory could have appeared a compromise between the pure field theories of Maxwell and Hertz, and the particulate theories of Weber and Helmholtz. But it was a rather elaborated theory, although it was soon superseded by Einstein's relativistic Electrodynamics.

The conceptual relevance of theoretical debates among three great expert of the end of century Electrodynamics, however, cannot ignore or understate the sometime decisive role of experiments in their role of deciding between alternative positions. I believe that the most important piece of information for the construction of microphysics, alternative to Maxwell's and Hertz's, and to Lorentz' theory, was Pieter Zeeman's amazing prediction in August 1896 that the oscillating particles emitting light had a subatomic mass. An historical and epistemological interest is therefore presented by Zeeman 1902 Nobel Lecture description of the story of his discovery of the Zeeman effect[35]. As Zeeman tells in his Lecture, on the story of his first approach to the 1896 experiment, he was trying to obtain the effect just by relying on the greater resolution of his spectroscope, a Rowland concave reflection grating, improved by screening the source with a slit. In spite of these instrumental improvements, the result was very uncertain because, after the magnet was powdered, the sodium line was not split but just slightly widened (two tenths of an Angstrom) and this effect could not be unambiguously explained in terms of the magnetic force because thermodynamic effect could also have contributed to it. Zeeman was very dubious of the interpretation of the result since Maxwell's authoritative pronouncement against a positive result was in the back of his mind. But he changed his expectations on the splitting of the lines, explained it through Lorentz's particulate hypothesis, confirmed the predicted new type of splitting (a doublet) in the direction of the magnetic force, and checked that their polarization was as the theory predicted. Perhaps the most important piece of information for the construction of microphysics was Zeeman's amazing prediction in August 1896 that the oscillating particles emitting light had a subatomic mass. He thus strongly contributed to the discovery of the first subatomic particle. In 1902 Zeeman received with Lorentz his Nobel Prize.

6. EINSTEIN'S RELATIVITY: IS ETHER A PROPAGATION MEDIUM FOR THE VELOCITY OF LIGHT?

Following my reading of E.'s original contributions, I believe that the popular understanding of E.'s decision to abolish ether in his celebrated 1905 essay on special relativity needs deeper historical and philosophical analysis. E.'s ideas on the ether problem are in fact strictly related to his technical and philosophical approaches to the whole context of his revolutionary relativity theories. E.'s original ideas on ether are condensed in one of his important, but less quoted essay, devoted to his early discussions on ether in his 1920 book Relativity, the Special and General Theory[36], and to his return to the same problem in his 1954 fifth Appendix added to the book, and titled: "Ether and the Theory of Relativity and the Problem of Space". Let us resume E.'s most interesting remarks on ether in his 1920 publication. The "far-reaching similarity, which subsists between the properties of light and those of elastic waves in ponderable bodies", represented in E.'s view "a fresh support", for an elastic type of ether as a medium for Light's waves. But the elastic approach to theory was also a source of great difficulties, because "neither Maxwell nor his followers succeeded in elaborating a mechanical model for the ether which might furnish a satisfactory mechanical interpretation of Maxwell's laws of the electro-magnetic field. The laws were clear and simple, the mechanical interpretations clumsy and contradictory" (My Italics)[37]. In E.'s view, his special theory of relativity overcame the difficulties by an hard restriction. In his essay he assumed that ether consisted of particles whose motion was not observable through time: "There may be supposed to be extended physical objects to which the idea of motion cannot be applied... The special theory of relativity abstracted from ether the last mechanical characteristic. But, a non mechanical ether might appear as a superfluous requirement in place of electromagnetic fields as ultimate, irreducible realities...If from the standpoint of ether this hypothesis appears at first to be an empty hypothesis, one should consider that in the electromagnetic processes in vacuo... the electromagnetic fields appear as ultimate, irreducible realities, and at first it seems superfluous to postulate a homogeneous, isotropic ether-medium, and to envisage electromagnetic fields as states of this medium". As unexpected as it might seem, this hypothesis aims to reconcile the two contrasting position of a mechanical or an electromagnetic state of ether, because Einstein does not renounce to ether: "on the other hand there is a weighty argument to be adduced in favor of the ether hypothesis. To deny the ether is ultimately to assume that empty space has no physical qualities whatever. The fundamental facts of mechanics do not harmonize with this view". Surprising enough, Newton's authority is still called in support of a physical conception of space. According to Einstein in the general tendency to give conceptual priority to electromagnetic concepts sought a solution to the ether problem, by reducing the principles of mechanics to those of electricity: "especially as a confidence in the strict validity of the equations of Newton's mechanics was shaken by the experiments with radioactive-rays and rapid cathode rays". A positive influence for this solution, represented by the electro-dynamical investigations of Heinrich Hertz, was neglected by Einstein because "the dualism still confronts us... in the theory of Hertz, where matter appears not only as the bearer of velocities, kinetic energy, and mechanical pressures, but also as the bearer of electromagnetic fields".

E.'s approach to ether couldn't disregard his relation with Lorentz, the authoritative supporter of an ethereal reality. Einstein tends to conciliate the contrasting feature of the Lorentian approach: "The space-time theory and the kinematics of the special theory of relativity were modeled on the Maxwell-Lorentz theory of the electromagnetic field". Moreover: "According to Lorentz's theory, electromagnetic radiation, like ponderable matter, brings impulse and energy with it, and as, according to the special theory of relativity, both matter and radiation are but special forms of distributed energy, ponderable mass losing its isolation and appearing as a special form of energy"[38].

The role of a new ether was part of E.'s life interest, and of his recurrent hope of an unification of electromagnetism and gravitation. Einstein's remarks in the same essay are thus presented: "As to the part which the new ether is to play in the physics of the future we are not yet clear. We know that it determines the metrical relations in the space-time continuum, e.g., the configurative possibilities of solid bodies as well as the gravitational fields; but we do not know whether it has an essential share in the structure of the electrical elementary particles constituting matter. Nor do we know whether it is only in the proximity of ponderable masses that its structure differs essentially from that of the

Lorentzian ether; whether the geometry of spaces of cosmic extent is approximately Euclidean".

In his General Relativity in fact ether was accepted, as long as its physical nature could be conceived as something different from the centuries old concept of an uniform distributed special substance: "What is fundamentally new in the ether of the general theory of relativity as opposed to the ether of Lorentz consists in this, that the state of the former is at every place determined by connections with the matter and the state of the ether in neighboring places, which are amenable to law in the form of differential equations; whereas the state of the Lorentzian ether in the absence of electromagnetic fields is conditioned by nothing outside itself, and is everywhere the same. The ether of the general theory of relativity is transmuted conceptually into the ether of Lorentz if we substitute constants for the functions of space, which describe the former, disregarding the causes, which condition its state. Thus we may also say I think, that the ether of the general theory of relativity is the outcome of the Lorentzian ether, through relativation" (Italics S.D).

The theme of a theory that could succeed in associating gravitational and electromagnetic field as one unified conformation, interested Einstein's research on a Generalized Field Theory for the rest of his life. In this connection, E.'s early neglect of ether declines in his mature reflections, and more comprehensive considerations are consistent with the fundamental tenets of his general relativity: "More careful reflection teaches us however, that the special theory of relativity does not compel us to deny ether. We may assume the existence of an ether; only we must give up ascribing a definite state of motion to it, i.e., we must by abstraction take from it the last mechanical characteristic which Lorentz had still left it. We shall see later that this point of view... is justified by the results of the general theory of relativity". This view of his ether is to be regarded as his last conviction in the closing years of his life: "Recapitulating, we may say that according to the general theory of relativity space is endowed with physical qualities; in this sense, therefore, there exists an ether. According to the general theory of relativity space without ether is unthinkable; for in such space there not only would be no propagation of light, but also no possibility of existence for standards of space and time (measuring-rods and clocks), nor therefore any space-time intervals in the physical sense. But this ether may not be thought of as endowed with the quality characteristic of ponderable media, as consisting of parts which may be tracked through time. The idea of motion may not be applied to it."[39]

7. CONCLUDING REMARKS

The problem of the nature of Light and of the role of Ether, as its medium or source, was very central in the scientific literature at the end of the Nineteenth Century, and thus one may take account for the German and European scientists' reaction to E.'s abolition in his 1905 essay. A short and approximate list of his intellectual opponents cannot miss to mention Mach's indifference, Poincaré and the French culture silence, not to mention the harsh and partially motivated criticism of colleagues in E.'s own country. Following J.J. Thomson's discovery, physicist considered the oscillating electrons as the Light's physical source. Were electrons performing the role of the old-fashioned Keplerian Lumen? The demand can evidently be answered on a physical level, either on a cosmological basis, as by a Big Bang type modern theory, or by the black matter mysterious presence. In this paper, I argued, however, that such questions also deserved an historical and philosophical approach. As for an example, although Helmholtz' and Maxwell's theories were conditioned by their technical expertise, their arguments concerned the philosophical problem of perceptions and its bearing on the nature of Light. Let us consider in fact Hertz's case on the discovery of Maxwell's waves. I noticed a remarkable interdependence between the final Hertz's representation of the waves and the philosophical conception of ether that was a background to his discovery. The discovery corresponded to the moment when he realized that the conception of a dielectric "a la Poisson" was not sufficient to support the conviction of the existence of waves of electric force in air, but that their existence had to be conceived as a primitive polarization of ether. The discovery was in some way a consequence of this change of conceptions and strengthened, in turn, this change. I consider this example as a very relevant instance of the strength of preconceived, i.e., a-priori ideas in physics research.

Although the study of Light limited to its visual properties represented the bases themselves of the initial science of optics, and of the nature of Light, the discovery of electromagnetic non optical of X rays had the rather anomalous effect of diminishing the intrinsic interest for the nature of Light. It was replaced by an increased research on similarities within the extended band of electromagnetic waves, whose whole spectrum covered several millions of units Hertzes, whereas the visible section included just three hundreds Hertzcs.

No wonder that in this situation Helmholtz represented one of the leaders in middle of Century in physical and physiological optics. By his regularity Principle he hardly tried to discriminate between physical and physiological principles. In my view, his thought stands on the watershed between two great traditions in the development of modern physics. On the one hand, through the attention given to perceptions per se and psychology, he opened the way to Mach's phenomenology. On the other hand, with his special upholding of Kant's a-priori, he favored Hertz's philosophy and Hertz's particular re-evaluation of Kant's a-priori.

Passionate discussions among physicists were very frequent in the same period of Einstein's intervention. But the historian Rosenfeld rightly remarked that one of the effects of Hertz's great experiment was to deviate the major line of discussion from the ether's to the problem of the evident similarity between the well known properties of Light, and that of the recently produced electromagnetic waves of various frequencies (X rays included). Their experimental production had the effect of polarizing the physicist's interest, relinquishing to a secondary level the problem of their nature and of their relation with ether[40].

On the other side, the international success of Maxwell's and Hertz's local field theories, and the almost cotemporary triumph of Lorentz's and Zeeman's antagonist theories of an atomistic microphysics, required a new type of analysis to scientists, historians, and philosophers of science. Lorentz, the best representative of the recently developed particles and waves Electrodynamics, frequently expressed reactions to the ether abolition. Although ontological determinism, realism, and purely phenomenological positions, were "a la pair", Einstein did not regard his decision as one of his alleged antirealist approaches to physics[41].

It was Einstein's merit in 1905 to perceive that excluding Ether in his Special Relativity would advantage theoretical physics. However, he did not consider its exclusion as an apodictic decision. At the end of his scientific life he was still fundamentally convinced of the indispensability: "According to the general theory of relativity space without Ether is unthinkable: for in such a space there would not only be no propagation of light but also no possibility of existence for standards of space and time"[42]. E. never lost faith in a conciliation of GR with Quantum Theories. Let us consider his conclusive statement: "Further, in contemplating the immediate future of theoretical physics we ought not unconditionally to reject the possibility that the facts comprised in the quantum theory may set bounds to the field theory beyond which it cannot pass".

Let us conclude: E.'s decision in 1905 to abolish ether in his special relativity did not settle the problem once for all. Let me try to condense in a few drastic sentences Einstein's alleged doubts in front of his dramatic decision. If one accepts (pace Maxwell and Hertz) that Light is electricity, and that ether is analogous to electricity itself (pace Hertz's dielectric ether), how might one abolish ether without abolishing electricity? Einstein's solution consisted in extracting from ether the old fashioned ethereal attributes, of substance and motion, and leaving apart a few properties listed in his considerations. It is thus comprehensible that he often returned to arguments concerning his logical and epistemological justification. They were variously motivated, but on the whole, let me allow to drastically resuming two Einstein's divergent positions. Accepting the role of ether as a material medium for transmission of electromagnetic waves and of Light presented the great difficulty of theoretically justifying the enormous high value of Light's velocity, whereas any attempt at an elastic type form of mechanical explanation was by him definitively rejected. On the other side, as shown above, he was reluctant to view the magnetic and electric fields as entities independently existing and moving with a specific velocity in a vacuum space-time. From an historical perspective, the Einstein's time conceptual relationship among ether, Light, and elctromagnetic waves, was very far from presenting hopes for an axiomatic rational approach. Einstein settled the problem in his General Relativity by stating that in line of principle the velocity of Light was a universal constant of nature.

REFERENCES

1. Bartellini, F. (2010). La scomparsa della sfera del fuoco ela nascita del fuoco d'amore. Novi Ligure: Edizioni Joker.
2. Cassirer, E. (1950). The Problem of Knowledge, Philosophy of Science and History of Science since Hegel. New Haven, CT: Yale University Press.
3. D'Agostino, S. (1975). Hertz's Researchers on Electromagnetic Waves. Historical Studies in the Physical Sciences, 6, 261-323.
4. D'Agostino, S. (1996). Absolute Systems of Units and Dimensions of Physical Quantities. Physis (Rivista Internazionale di Storia della Scienza), XXXIII, Fasc. 1-3, NS.
5. Einstein, A. (1922). Ether and Relativity. Sidelights on Relativity. Trans by Lawson, R. W. London: Henry Holt and Company.
6. Hendry, J. (1986). James Clerk Maxwell and the Theory of Electroamgnetic Field. Bristol and Boston: Adam Hilger Ltd.
7. Hermann von Helmholtz (1977). Epistemological Writings: The Paul Hertz/Moritz Schlick Centenary Edition of 1921, with Notes and Commentary by the Editor (Ed. with Introd. and

Bibliography by R. S. Cohen, & Y. Elkana). Boston Studies of the Philosophy of Science, Vol. 37. Dordrecht: Springer.

8. Hertz, H. (1956a).The Principles of Mechanics Presented in a New Form. Prefase of H.von Helmholtz.Autorized English Translation by Jones D.R. and Walley J.T. with a new Introduction by Cohen R.S. New York: Dover Pbl. Inc.

9. Hertz, H. (1956b). On the Relation between Maxwell's Fundamental Electromagnetic Equations and the Fundamental Equations of the Opposing Electromagnetics. Miscellaneous Papers. London: Macmillan. English Translation of Hertz, H. (1884) Schriften vermischten Inhalts.Gesammelte Werke, Band I. Leipzig: J.A. BarthVerlag.

10. Hertz, H. (1962). Electric Waves, Being Researches on the Propagation of Electric Action with Finite Velocity through Space. New York: Dover Pbl Inc. Authorized English Translation by Jones D.E. of Hertz, H. (1892). Untersuchungen ueber die Ausbreitung der elektrischen Kraft. Leipzig: Johann Ambrosius Barth.

11. Howard, D. (1993). Was Einstein Really a Realist? Perspectives on Science: Historical, Philosophical, Social, 1, 204-251.

12. Hunt, B. J. (1991) The Maxwellians. Cornell History of Science Series.

13. Lorentz, H. A. (1952). Poblems of Modern Physics. New York: Dover.

14. Maxwell, J. C. (1954) The Scientific Papers of James Clerk Maxewll. Two Volumes. New York: Dover Publications.

15. Planck, M. (1931). James Clerk Maxwell in seine Bedueting fuer die Theoretische Physik in Deutschland. Naturwiss, 1889- 1894.

16. Ronchi, V. (1952). Storia della Luce. Fondazione "Giorgio Ronchi". Second Edit. Zanichelli, Bologna: Casa Edit.

17. Rosenfeld, L. (1956). The Velocity of Light and the Evolution of Electrodynamics. Il Nuovo Cimento, 4, 1630-1667.

18. Zeeman, P. (1967) Light Radiation in a Magnetic Field. Nobel Lectures. Physics 1901-1921. Amsterdam: Elzevier Publ. Co.

Chapter 10

Electromagnetic Wave Propagation in Body Area Networks Using the Finite-Difference-Time-Domain Method

Jonathan N. Bringuier [1,*] **and Raj Mittra** [1,2]

[1] *Department of Electrical Engineering, The Pennsylvania State University, University Park, PA 16802, USA*

[2] *King Fahd University of Petroleum and Minerals, Dhahran 31932, Saudi Arabia*

ABSTRACT

A rigorous full-wave solution, via the Finite-Difference-Time-Domain (FDTD) method, is performed in an attempt to obtain realistic communication channel models for on-body wireless transmission in Body-Area-Networks (BANs), which are local data networks using the human body as a propagation medium. The problem of modeling the coupling between body mounted antennas is often not amenable to attack by hybrid techniques owing to the complex nature of the human body. For instance, the time-domain Green's function approach becomes more involved when the antennas are not conformal. Furthermore, the human body is irregular in shape and has dispersion properties that are unique. One consequence of this is that we must resort to modeling the antenna network mounted on the body in its entirety, and the number of degrees of freedom (DoFs) can be on the order of billions. Even so, this type of problem can still be modeled by employing a parallel version of the FDTD algorithm running on a cluster. Lastly, we note that the results of rigorous simulation of BANs can serve as benchmarks for comparison with the abundance of measurement data.

Keywords: Body Area Networks; FDTD; conformal; numerical phantoms; Cole-Cole model; recursive convolution method; mutual coupling

1. INTRODUCTION

As wireless technologies continue to evolve, the development of personalized devices which exchange data with unprecedented ease and efficiency is certain to grow unabated. Recently, such technologies have garnered much attention from those interested in biomedical sensing and Body-Area-Networks (BANs). These technologies, which are data networks that operate through wireless transmission using the human body as a propagation medium, continue to be an active area of research in fields such as remote health monitoring and medical biosensors. However, the complexity of integrating efficient communication systems in the human body environment poses many challenges to the future of this emerging technology. The excitation of surface and space waves from radiating antennas mounted on the body can have a large impact on the performance of co-site body-centric antenna systems. Moreover, the presence of multiple antennas on the body can lead to unexpected coupling due to creeping wave interaction. Fortunately, recent developments in powerful numerical methods, such as the Finite-Difference-Time-Domain (FDTD) method running on parallel platforms, has made it feasible for us to carry out a detailed study of these body-centric antenna systems.

2. SIMPLE MODELS FOR BANS

For many years researchers have used simplified geometries to model the interaction of electromagnetic energy with biological tissue. Such geometries typically lend themselves to simple shapes, such as cylinders, ellipses and spheres for which past research has typically treated these as perfectly conducting objects, low loss high dielectrics, and surface impedance models. This type of treatment was convenient due to the availability of well-known analytical methods used solve these problems, such as the uniform theory of diffraction (UTD), ray tracing (RT), creeping waves, and eigenfunction analysis [1,2]. Indeed, these models have been shown in recent years to yield reasonably good results for simple cases that compare well to those derived numerically rigorously with modern CEM techniques. However, the asymptotic techniques fail to perform well when the dielectric medium they deal with is arbitrarily shaped, inhomogeneous and lossy, and such problems must be handled by using general-purpose numerical methods, such as the FDTD [3–6]. Nevertheless, these simple geometries can still provide a computationally efficient way to study the propagation around and through the human body. Therefore, these models remain useful for studying BANs.

In this section we provide the results of a simplified model for the human body torso. The proposed model is a 3-layer elliptical structure having major and minor axis of 150 cm and 120 cm, respectively. This model has been used in [7] to investigate the coupling around a 2-D ellipse using the sub-band FDTD, UTD/RT and measurement techniques at UWB frequencies. However, in that work the model was assumed to be perfectly conducting when applying the UTD/RT method while a homogeneous muscle phantom was employed for the sub-band FDTD analysis. Additionally, a conformal FDTD algorithm was not used to improve the accuracy of the curved surface whereas the author has done so in this work. Furthermore, it neither accounted for the multi-layer dielectric properties of the tissue in the human torso, nor did it simulate the actual radiating element, i.e., viz., the monopole antenna rigorously.

The 3-layer ellipse model incorporates the skin, fat, and muscle layers. In [8] a simple 3-layer planar slab model using 3 mm skin layer, 5 mm fat layer and muscle was used to study the penetration depth of an incident plane wave for use in implantable medical devices. In this work we have used the same thickness for the skin and fat layers in the elliptical model.

Although the FDTD method enjoys a significant advantage over MoM in terms of its ability to simulate complex structures and lossy inhomogeneous materials, it has been known to require far more computational resources than are usually available to accurately model, simultaneously, the fine features of the radiating element, the layered structure of the geometry and the electrically large size of the entire structure typically encountered in the study of BANs. To circumvent this problem, most research done on BANs by utilizing the FDTD have been carried out using a point source approximation. While such an approximation may be adequate for directly estimating the path loss associated with the body, it does not rigorously account for a number of crucial antenna factors that affect the antenna performance, such as finite ground size, radiation pattern and efficiency. Furthermore, any field behavior in either the near or the intermediate region is inherently neglected in the point source approximation, whereas the physical structure of the radiating element must be included to properly model the physical system.

In this paper we have used a parallel version of FDTD that can handle such electrically large geometries as well as the fine features of the radiating element. To this end, we have used quarter-wave monopoles resonant at 2.45 GHz with a 75 mm × 75 mm square ground plane located at tangent points one to two FDTD cells away from the models analyzed in this study. Although other antennas could be used, the choice of the monopole antenna has the distinct advantage in that the radiation pattern in the plane azimuthal to the monopole is inherently stable across the bandwidth of interest. However, the monopoles used in this study do not have an impedance matching bandwidth wide enough to cover the frequencies of interest, and, therefore, the channel path loss data between observation points should be considered relative to one another for a given frequency. Nevertheless, the polarization and radiation pattern deviations of the transmitter and receivers across the frequency band can more or less be neglected when performing coupling calculations, which make these antennas useful for studying channel characteristics of the on-body propagation medium.

The setup used in the simulation is shown in Figure 1 where the receiver is located first in the source plane and then displaced vertically by 210 mm and 400 mm, respectively, from the above plane. For each observation plane the receiver is moved around the elliptical trunk model and the S_{21} is recorded. The mode of propagation is known to be a creeping wave and the direct ray paths are shown. For each observation plane the S_{21} was calculated along the elliptical path in the level plane and plotted for frequencies between 0.8–6 GHz as shown in Figures 2–4.

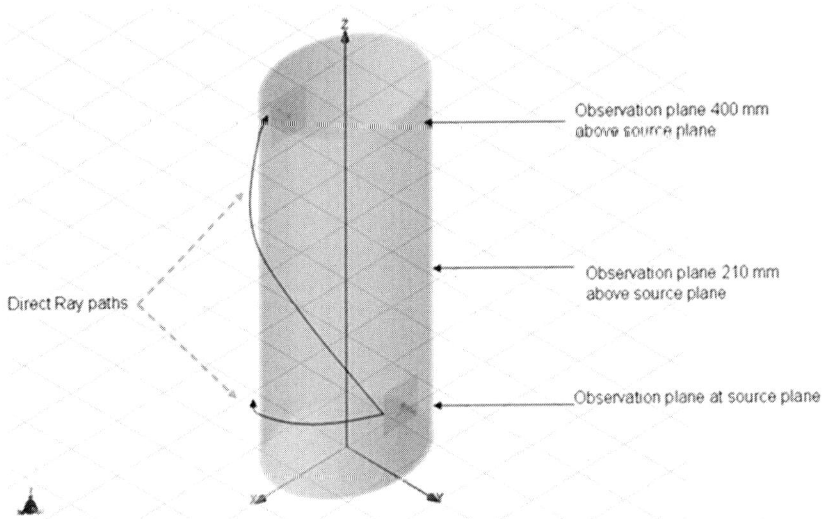

Figure 1. 3-layer ellipse model of the human torso with transmitting antenna at the front and receiving antenna at the back.

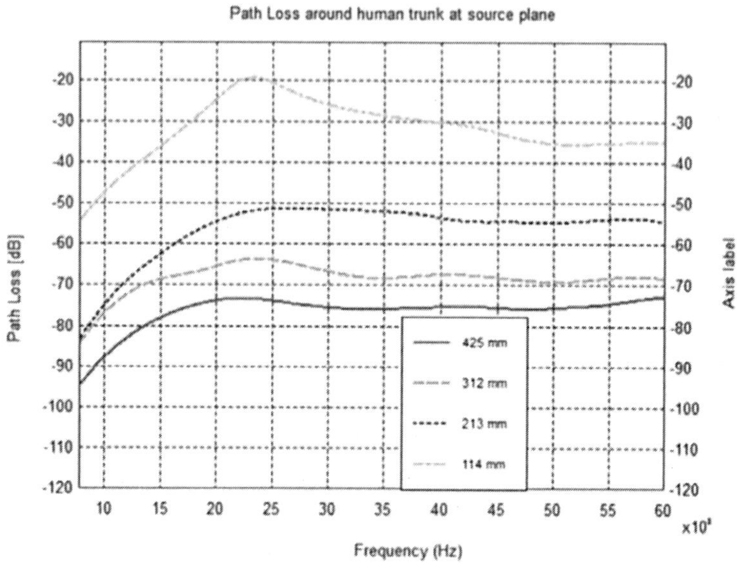

Figure 2. Path loss around the cylindrical human trunk model at the source plane.

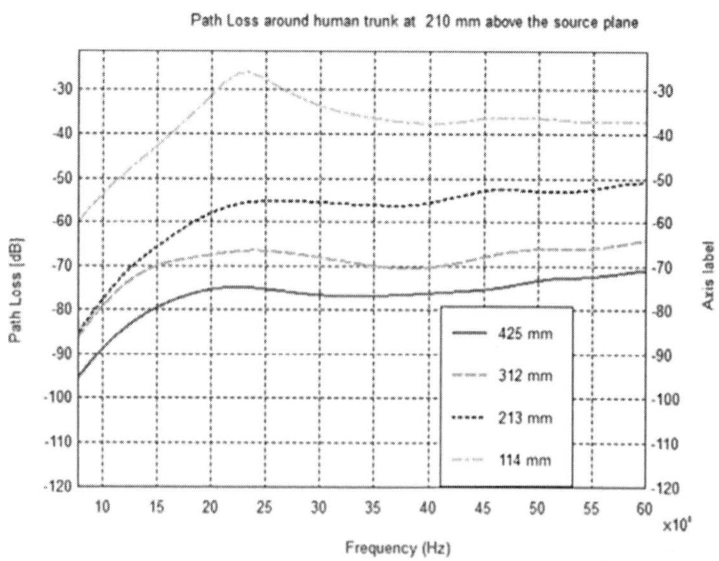

Figure 3. Path loss around the cylindrical human trunk model 210 mm above source plane.

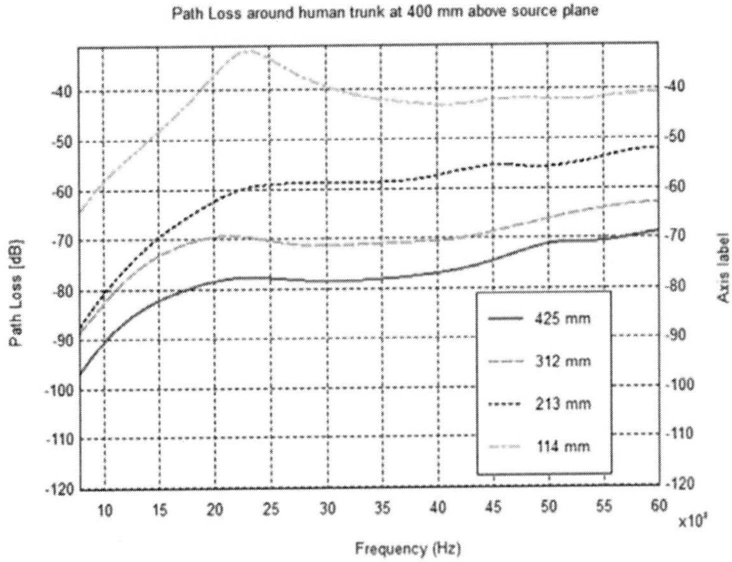

Figure 4. Path loss around the cylindrical human trunk model 400 mm above source plane.

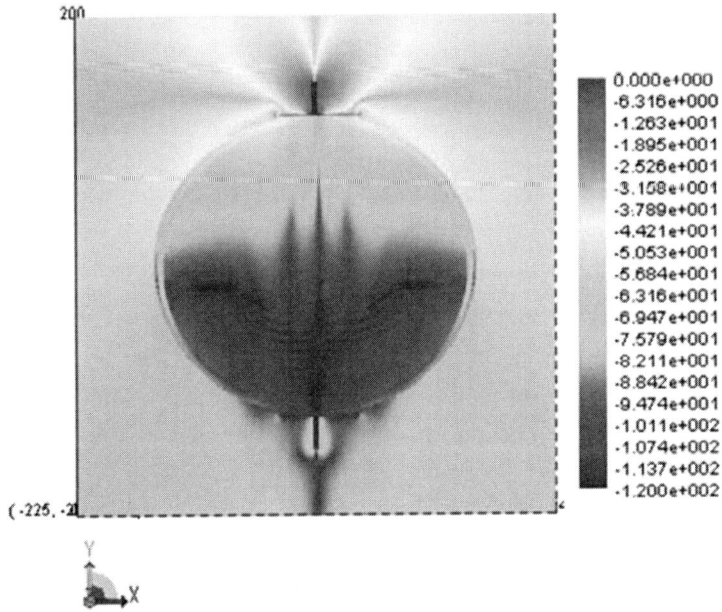

Figure 5. Electric field distribution in the source plane.

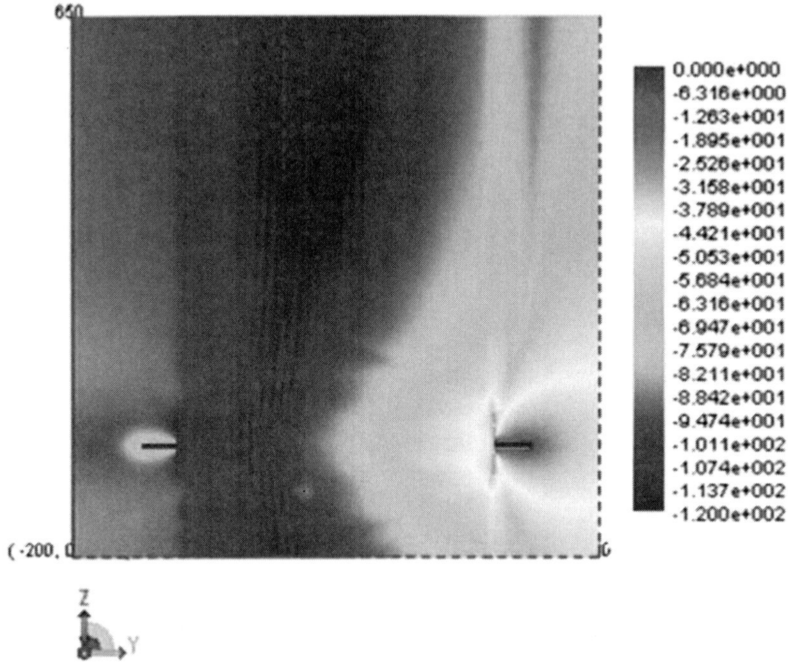

Figure 6. Electric field distribution on a vertical cut plane bisecting the cylindrical model.

For BANs it is typically assumed that the wave propagation through the body is so attenuated that its contribution can be ignored. To verify this we have plotted the fields for the cross-section of the ellipse in the source plane as well as a vertical cross-section along the extent of the ellipse model. The electric field plots are shown in Figures 5 and 6. From these figures it can be seen that the field magnitude inside the model is less than −100 dB and the main source of energy transfer is along the surface of the trunk in the form of a creeping wave.

In Figures 7–9 the path loss is plotted versus the distance around the elliptical model in each of the different observation planes. As shown in Figure 7, which depicts the observations made in the source plane, the attenuation at higher frequencies exhibit a clear monotonic behavior and is characteristic of a creeping wave. The results in Figures 7–9 have been found to agree closely with those in the literature for other body

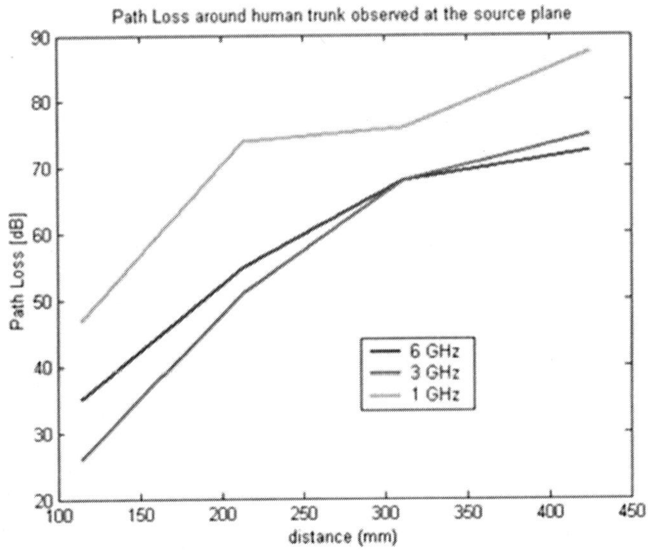

Figure 7. Path loss versus separation distance with receiving antenna at the source plane.

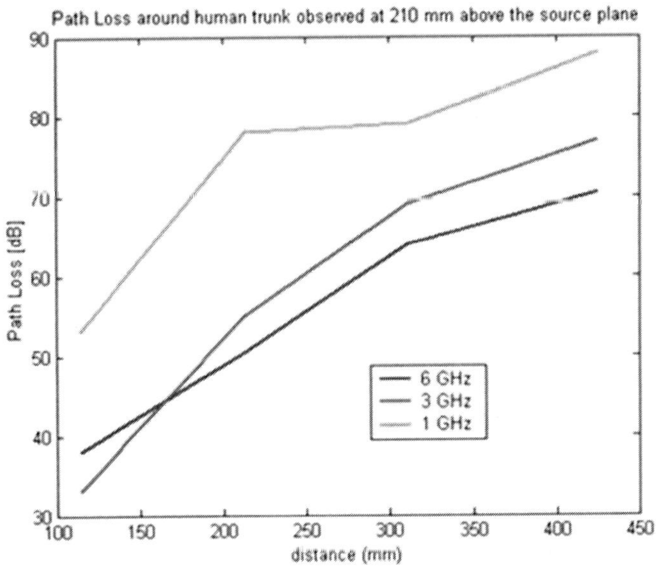

Figure 8. Path loss versus separation distance with receiving antenna 210 mm above the source plane.

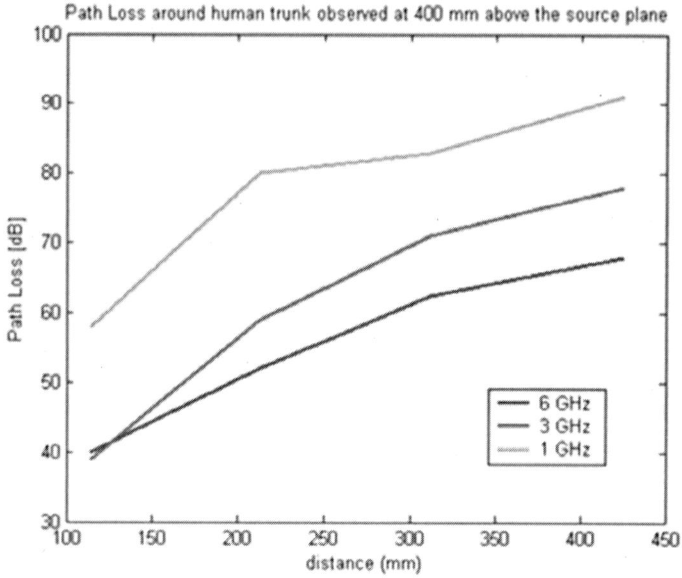

Figure 9. Path loss versus separation distance with receiving antenna 400 mm above the source plane.

models [1,2,7,9–19]. These results can be useful as reference points when approximating the rate of attenuation along the trunk in body-centric network scenarios.

3. NUMERICAL PHANTOMS FOR BANS

Although simplified geometries have their use in analyzing body-centric communication networks, the availability of realistic human body models enables us to use a more rigorous approach to understand how the irregular shape of the body affects the performance of BANs. One of the most commonly used body models is shown in Figure 10, where the original data set is comprised of voxels (i.e., 3D matrix data set). Furthermore, a hypothetical BAN configuration is shown with the transmitting and receiving antennas. To reduce the computational resources needed in the simulation, a lossy homogeneous model was used.

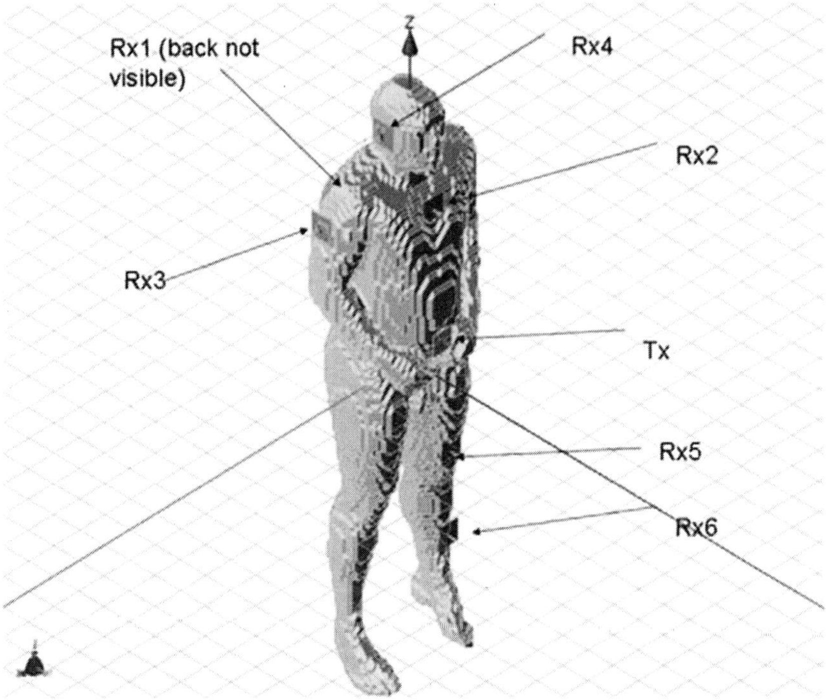

Figure 10. Numerical human body phantom based on 3D CT scan voxel set with transmitting and receiving antennas in typical BAN scenario.

Additionally, the resolution of the original voxel data set, which is 1 mm × 1 mm × 1 mm, was down-sampled to 3 mm × 3 mm × 3 mm. Even at this lower resolution the computational resources needed are high, specifically 8 hours on 16 Pentium 4 CPUs. Although this may seem too coarse for the conventional $\lambda/20$ cell size, the fields couple on or near the surface of the body, where the mesh can be coarser, and we are interested in separation distance well into the asymptotic region of the transmitting antenna. To demonstrate that the loss of accuracy is negligible in doing so, we have performed a simple experiment by computing the difference in field magnitude at 3 GHz—far away from the transmitting antenna on the same plane as shown in Figure 11. It is evident from the results that in the asymptotic region there is little loss in accuracy by utilizing a down-sampled phantom model.

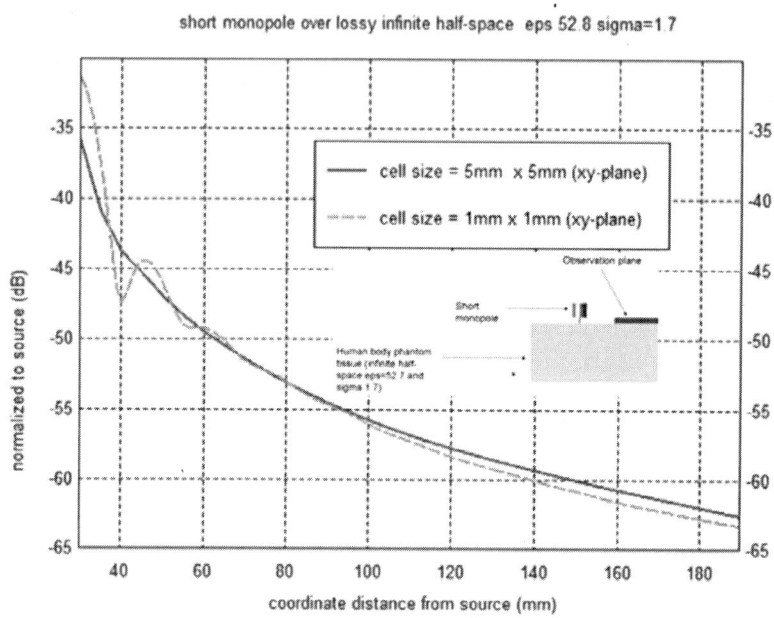

Figure 11. Experiment to determine if the down-sampled human body voxel set causes numerical inaccuracies (λ = 10 cm).

4. DIELECTRIC PROPERTIES AND DISPERSION MODELS OF BIOLOGICAL TISSUE

It is commonly claimed in the literature that a good approximation for the body can be made by using a full muscle phantom or a two-thirds muscle equivalent phantom, the latter being more commonly used in EMC analysis for SAR [20,21]. However, realistic biological tissues exhibit dispersion characteristics that are not of the traditional Debye type used in the FDTD model, whose spectral behavior is usually represented by the form:

$$\hat{\varepsilon}(\omega) = \varepsilon_{\infty} + \sum_{n=1}^{N} \frac{\Delta\varepsilon_n}{1 + j\omega\tau_n} + \frac{\sigma_n}{j\omega\varepsilon_0} \quad (1)$$

where τ_n is the relaxation time constant $\Delta\varepsilon_n$ is the spectral coefficient and σ_n is the static ionic conductivity. This model can be handled in FDTD through a simple convolution extension of the basic FDTD algorithm. In general, biological tissue exhibit a broadened spectrum which is empirically accounted for by the parameter $0 \leq \alpha_n < 1$, (see Gabriel et al. [22]) and is typically referred to as the Cole-Cole model. It is given by:

$$\hat{\varepsilon}(\omega) = \varepsilon_\infty + \sum_{n=1}^{N} \frac{\Delta\varepsilon_n}{1 + (j\omega\tau_n)^{1-\alpha_n}} + \frac{\sigma_n}{j\omega\varepsilon_0} \tag{2}$$

The difficulty in this model lies in the fact that there are not any simple closed-form expressions amenable to the FDTD algorithm. Specifically, this is due to the fact that the inverse Fourier transform is non-trivial.

4.1. Dielectic Spectrum Approximation

In this Section we will introduce a technique that can be used to implement the Cole-Cole model into the desired frequency band of interest for simple geometries relevant to BANs (e.g., planar human tissue phantoms) in FDTD. Since the static conductivity term can be handled easily in the FDTD update equations we turn our attention to the term containing α_n. First we postulate that each term in the Cole-Cole model can be rewritten as an integral representation of a Debye spectrum as follows:

$$\frac{\hat{\varepsilon}_i(\omega)}{\Delta\varepsilon_i} = \frac{1}{1 + (j\omega)^{1-\alpha_i}} = \frac{1}{1 + (j\omega)^{\beta_i}} = \int_0^\infty \frac{g_i(\lambda)}{\lambda + j\omega} d\lambda \tag{3}$$

Then the expression for the electric field density follows as:

$$D(\omega) = \hat{\varepsilon}_i(\omega) E(\omega) \tag{4}$$

$$= \Delta\varepsilon_i \int_0^{\infty} \frac{E(\omega) g_i(\lambda)}{\lambda + j\omega} d\lambda \tag{5}$$

where by the inverse Fourier transform yields:

$$D(t) = F^{-1}\{\hat{\varepsilon}_i(\omega) E(\omega)\}(t) = \int_0^{\infty} E(t) \otimes e^{-\lambda t} g_i(\lambda) d\lambda \tag{6}$$

$$D(t) = E(t) \otimes L\{g_i(\lambda)\}(t) \tag{7}$$

Therefore, if it is possible to find a time domain representation for the Laplace transform of the unknown spectrum $g_i(\lambda)$ and, hence, we may extend the basic convolutional FDTD to implement non-Debye types of materials.

The problem at hand is to determine a suitable function for which we can derive the Laplace transform. To do this we first rewrite the Debye integral representation as:

$$\frac{\hat{\varepsilon}(\omega)}{\Delta\varepsilon} = \int_0^{\infty} \frac{g(\lambda)}{\lambda + j\omega} d\lambda = \int_0^{\infty} g(\lambda) \int_{-\infty}^{\infty} e^{-j\omega\rho} e^{-\lambda\rho} u(\rho) d\rho d\lambda \tag{8}$$

$$= \int_0^{\infty}\int_0^{\infty} g(\lambda) e^{-\lambda\rho} e^{-j\omega\rho} d\lambda d\rho \tag{9}$$

where u(ρ) is the Heaviside step function. Now by Fubini's Theorem:

$$\frac{\hat{\varepsilon}(\omega)}{\Delta\varepsilon} = \int_0^{\infty} e^{-j\omega\rho} \left[\int_0^{\infty} g(\lambda) e^{-\lambda\rho} d\lambda \right] d\rho \tag{10}$$

To find g(λ) we expand it in terms of some complete basis set for which we know the Laplace transform. Specifically, we chose Bessel functions of the first kind which form a complete set in $\Re$: $\lambda \in [0, \infty)$. Therefore, we have:

$$g(\lambda)=\sum_{n=0}^{\infty} C_n J_n(a\lambda) \tag{11}$$

where a is a fitting parameter. Obviously we must truncate this infinite series for numerical purposes; hence we write:

$$g(\lambda)\approx\sum_{n-0}^{N} C_n J_n(a\lambda) \tag{12}$$

Now we use the Laplace transform relation given by:

$$\int_0^{\infty} J_n(a\lambda)e^{-\lambda\rho}d\lambda=\frac{1}{\sqrt{\rho^2+a^2}}\frac{a^n}{\left(\rho+\sqrt{\rho^2+a^2}\right)^n} \tag{13}$$

Of course, we could have used Equation (1) directly in Equation (5); however, we have found that the integral converges much more slowly than that of the reformulation given in Equation (10). After inserting Equation (13) into the equation given in Equation (10) we have:

$$\frac{\hat{\varepsilon}(\omega)}{\Delta\varepsilon}=\int_0^{\infty} e^{-j\omega\rho}\left(\sum_{n=0}^{N} C_n\frac{1}{\sqrt{\rho^2+a^2}}\frac{a^n}{\left(\rho^2+\sqrt{\rho^2+a^2}\right)^n}\right)d\rho \tag{14}$$

Next, we can apply the method of least squares to find the coefficients C_n. However, from Equation (7) it is obvious that we must require the C_n's to be real. Therefore we have a constrained least squares system of equations given by the following:

$$\bar{b}^T=\left[\mathrm{Re}\int_0^{\infty}\frac{\hat{\varepsilon}(\omega)}{\Delta\varepsilon}I_0 d\omega \quad \mathrm{Im}\int_0^{\infty}\frac{\hat{\varepsilon}(\omega)}{\Delta\varepsilon}I_0 d\omega \quad \mathrm{Re}\int_0^{\infty}\frac{\hat{\varepsilon}(\omega)}{\Delta\varepsilon}I_1 d\omega \quad \mathrm{Im}\int_0^{\infty}\frac{\hat{\varepsilon}(\omega)}{\Delta\varepsilon}I_1 d\omega \quad \bullet \bullet \quad \mathrm{Im}\int_0^{\infty}\frac{\hat{\varepsilon}(\omega)}{\Delta\varepsilon}I_N d\omega\right. \tag{15}$$

$$\overline{x}^T = \begin{bmatrix} C_1 & C_2 & C_3 & \bullet & \bullet & \bullet & C_N \end{bmatrix} \tag{16}$$

$$\overline{\overline{A}} = \begin{bmatrix} \mathrm{Re}\int_0^\infty I_0 I_0 d\omega & \mathrm{Re}\int_0^\infty I_0 I_1 d\omega & \mathrm{Re}\int_0^\infty I_0 I_2 d\omega & \bullet & \bullet & \bullet & \mathrm{Re}\int_0^\infty I_0 I_N d\omega \\ \mathrm{Im}\int_0^\infty I_0 I_0 d\omega & \mathrm{Im}\int_0^\infty I_0 I_1 d\omega & \mathrm{Im}\int_0^\infty I_0 I_2 d\omega & \bullet & \bullet & \bullet & \mathrm{Im}\int_0^\infty I_0 I_N d\omega \\ \mathrm{Re}\int_0^\infty I_1 I_0 d\omega & \mathrm{Re}\int_0^\infty I_1 I_1 d\omega & \bullet & & & & \\ \mathrm{Im}\int_0^\infty I_1 I_0 d\omega & \mathrm{Im}\int_0^\infty I_1 I_1 d\omega & & \bullet & & & \\ \bullet & & & & \bullet & & \\ \mathrm{Re}\int_0^\infty I_N I_0 d\omega & & & & & \bullet & \\ \mathrm{Im}\int_0^\infty I_N I_0 d\omega & & & & & & \mathrm{Im}\int_0^\infty I_N I_N d \end{bmatrix} \tag{17}$$

where $\overline{\overline{A}}$ is 2NxN , $\overline{b}$ is 2Nx1 , $\overline{x}$ is Nx1 , $\overline{x} = \overline{\overline{A}}^{+}\overline{b}$ and

$$I_n = \int_0^\infty \frac{e^{-j\omega\tau} a^n}{\sqrt{\tau^2 + a^2} \cdot \left(\tau^2 + \sqrt{\tau^2 + a^2}\right)^n} d\tau \tag{18}$$

In common with most over-constrained inverse problems, we must use the Moore-Penrose pseudoinverse for $A^{\mp}$ to solve the system of equations. In practice, the integral cannot be evaluated over all possible frequencies; hence, it is typically replaced by a discrete summation over all frequencies of interest where the spectral match is desired.

As an illustrative example of this approach, the 4-term Cole-Cole model was taken for the tissue corresponding to the muscle. Although the preceding formulation corresponds to a single Cole-Cole term, the postulated Debye spectrum can equally well be applied to the entire summation of Cole-Cole terms. The parameters given by Gabriel [22] listed in Table 1.

Table 1. 4 term Cole-Cole model parameters for muscle.

$\Delta\varepsilon_1 = 50$	$\tau_1 = 7.23e-12$	$\alpha_1 = 0.1$
$\Delta\varepsilon_2 = 7000$	$\tau_2 = 353.68e-9$	$\alpha_2 = 0.1$
$\Delta\varepsilon_3 = 1.2e6$	$\tau_3 = 318.31e-6$	$\alpha_3 = 0.1$
$\Delta\varepsilon_4 = 2.5e7$	$\tau_4 = 2.27e-3$	$\alpha_4 = 0$

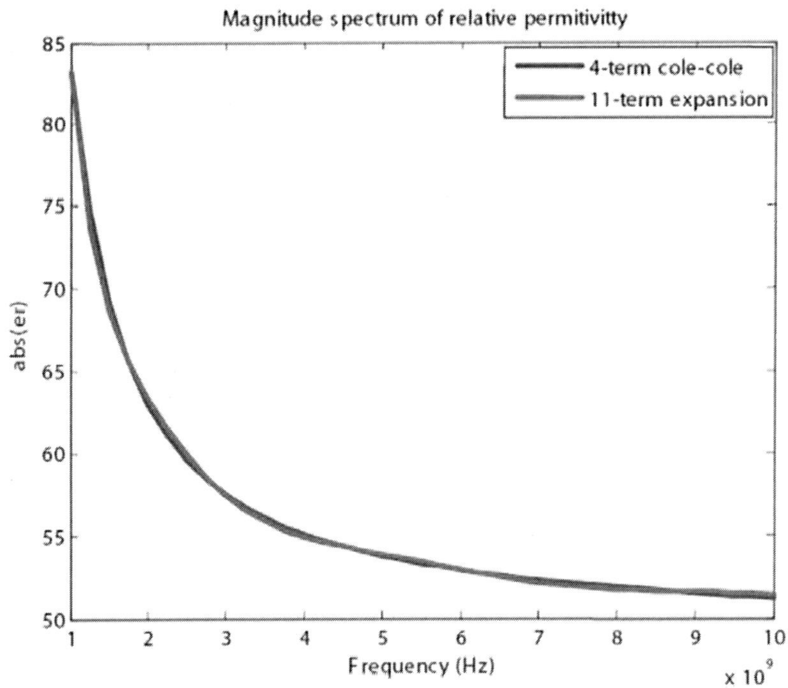

Figure 12. Magnitude spectrum of the relative permittivity for the 4 term Cole-Cole model and the spectrum approximation.

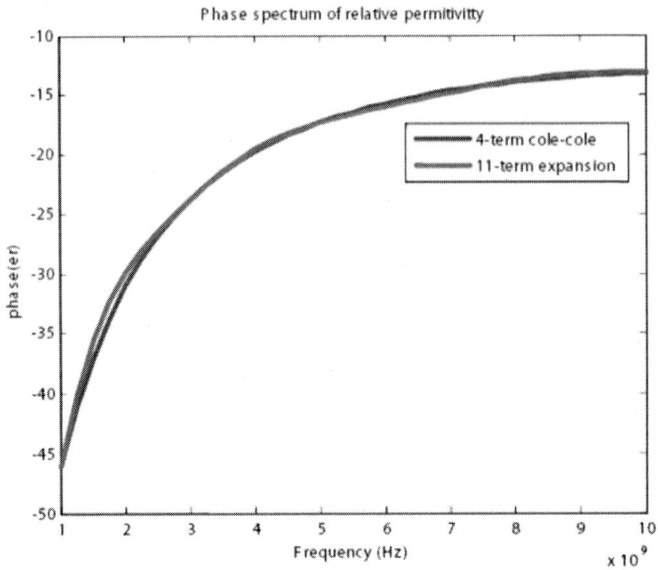

Figure 13. Phase spectrum of the relative permittivity for the 4 term Cole-Cole model and the spectrum approximation.

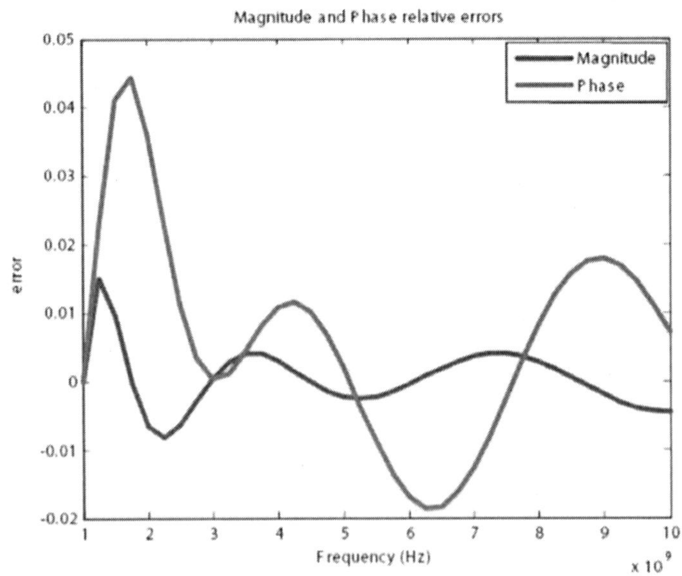

Figure 14. Magnitude and Phase relative errors of the spectrum approximation and the 4 Cole-Cole model.

Using the proposed algorithm we have obtained the following coefficients for an 11th-order expansion that are listed in Table 2. In addition, we have found that a good match was achieved when the fitting parameter was chosen to be a = 0.5e − 9. Although initially the order of the expansion may seem to be large, the expansion is only 2.75 times that of the original 4-term Cole-Cole model. Figures 12–13 show the magnitude and phase spectrum match for the band of interest, specifically 1–10 GHz. Furthermore, Figure 14 shows that the relative error is less than 5 percent over the entire band, for both magnitude and phase, and this can be considered a very good match for simulation purposes.

Table 2. Calculated coefficients from the spectral approximation method.

Coefficients	Numerical Value
C1	0.0002e6
C2	−0.0032e6
C3	0.0394e6
C4	−0.2759e6
C5	1.1027e6
C6	−2.4578e6
C7	2.7137e6
C8	−0.5489e6
C9	−1.7652e6
C10	1.6288e6
C11	−0.4334e6

As an example, the calculated coefficients along with the fitting parameter were used to model the frequency response in a muscle medium. Using the convolution sum of Equation (7) for a Gaussian pulse excitation in the medium yields the result shown in Figure 15. It can be seen from Figure 15 that the electric flux density inside the muscle medium is influenced by its frequency domain characteristics as modeled by the proposed algorithm.

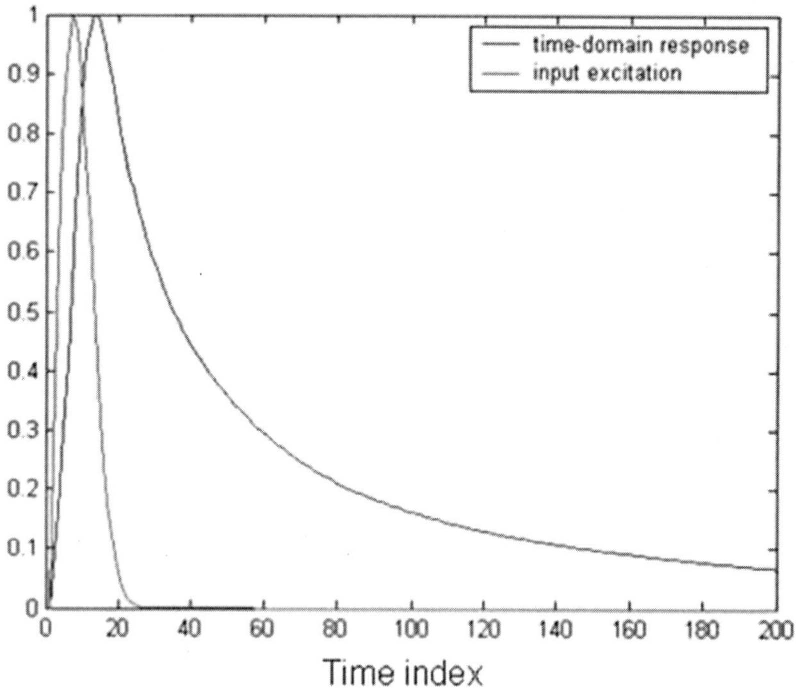

Figure 15. Normalized time-domain response of the electric flux density using the spectral approximation.

4.2. Recursive Convolution Method for Debye Materials

The problem with incorporating the spectral approximation algorithm proposed in the previous section is that FDTD requires a record-keeping of all past time values of the electric field components. If either the problem size is small, or reflection properties from an infinite planar slab are of interest then this approach may be feasible. However, for practical problems, the computational domain can become exceedingly large (e.g., upward of 1 billion unknowns for realistic human body phantoms), and other approximation methods must be used. Often times, the dispersion properties of biological tissue can be approximated by a Debye model over a finite frequency bandwidth. In this case it is possible to formulate an FDTD updating scheme for dispersive materials that does not require keeping all past histories of the fields.

The recursive convolution FDTD is one such method and is discussed in [5]. The essential features of this method are based on the Debye dispersion model given by:

$$\varepsilon(\omega) = \varepsilon_o \varepsilon_\infty + \varepsilon_o \chi(\omega) \tag{19}$$

where $\chi(\omega)$ is explicitly given as:

$$\chi(\omega) = \sum_{p=1}^{P} \frac{A_p}{\omega - W_p} \tag{20}$$

After some extensive manipulations the final form of the FDTD updating equations for the electric field become:

$$\vec{E}^{n+1} = \frac{\dfrac{\varepsilon_{eff}}{\Delta t} - \dfrac{\sigma_{eff}}{2}}{\dfrac{\varepsilon_{eff}}{\Delta t} + \dfrac{\sigma_{eff}}{2}} \vec{E}^{n} + \frac{1}{\dfrac{\varepsilon_{eff}}{\Delta t} + \dfrac{\sigma_{eff}}{2}} \vec{\nabla} \times \vec{H}^{n+\frac{1}{2}} + \frac{\dfrac{\varepsilon_o}{\Delta t}}{\dfrac{\varepsilon_{eff}}{\Delta t} + \dfrac{\sigma_{eff}}{2}} \sum_{p=1}^{P} \psi_p^n \tag{21}$$

and *ψnp* is given by:

$$\psi_p^n = \vec{E}^n \Delta \chi_p^0 + e^{jW_p \Delta t} \psi_p^{n-1} \tag{22}$$

The formulation given in Equation (21) can be used to incorporate materials that exhibit Debye-type dispersive properties into the FDTD. If the geometry is simple (e.g., a dispersive sphere), this method has been shown to give improved accuracy. However, in the case of the human body, where the object has irregular features, and many dispersive dielectric layers, the conformal (CFDTD) version, combined with the recursive convolution method, can lead to unstable results. In fact, when using the parameters in Table 1 with $\alpha_i = 0$ to simulate the body shown in Figure 10, the results were found to diverge. This problem further demonstrates the difficulties when trying to rigorously simulate a realistic human body. Therefore, we have used a common approximation [21] for the human body as consisting of a non-dispersive muscle ($\varepsilon_r = 52.7$, $\sigma = 1.7$) or its two-thirds equivalent phantom in the band of 0.8–3 GHz.

5. SIMULATIONS AND RESULTS

Using the muscle and the two-thirds muscle equivalent human body phantom, the CFDTD algorithm was used to simulate the setup previously shown in Figure 10.

For each model, the network configuration was simulated and the S_{21} was calculated. Additionally, the network was also simulated in the absence of the body in order to understand the influence of its presence. The free-space case (body absent) is shown in Figure 16 and the results of the two body models are shown in Figures 17 and 18. To gain insight into the field distribution around the body the electric field is shown in Figure 19 at 3 GHz. It is evident from the results that the line-of-sight (LOS) or near LOS paths are not greatly influenced by the presence of the body for monopole antennas.

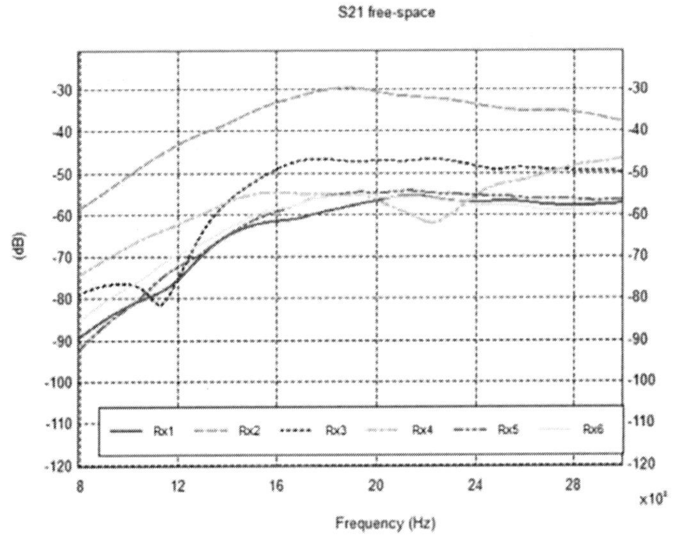

Figure 16. S_{21} of the BAN network scenario with body absent.

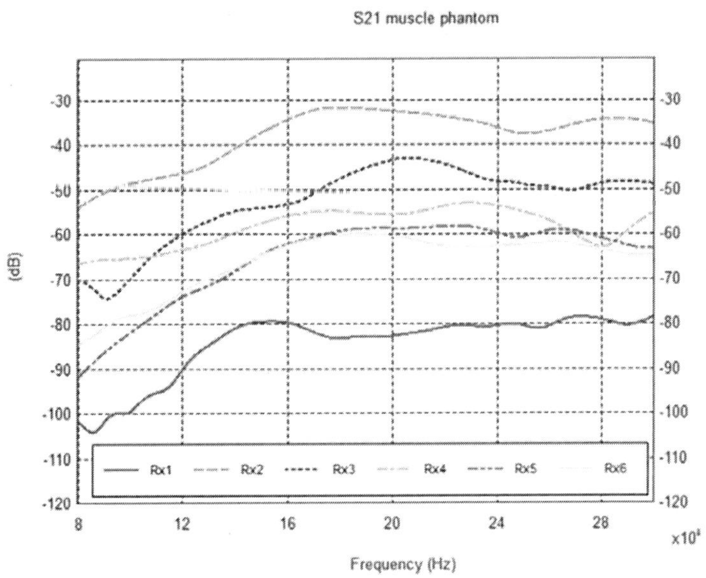

Figure 17. S_{21} of the BAN network scenario with muscle phantom.

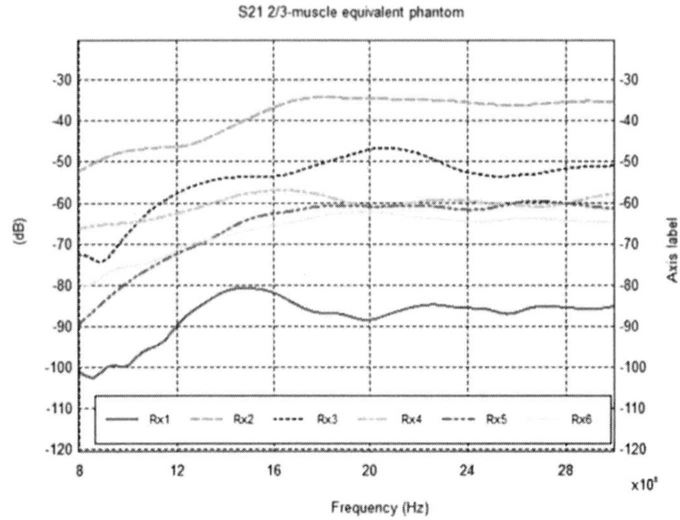

Figure 18. S_{21} of the BAN network scenario with 2/3-muscle phantom.

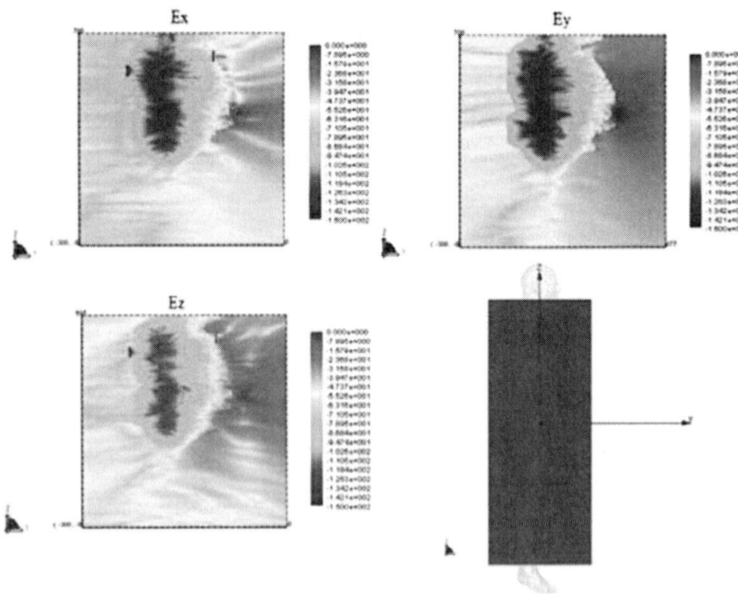

Figure 19. Electric field distributions (dB) on a vertical cut plane bisecting the body.

However, for the non-line-of-sight (NLOS) there is a noticeable difference in the path loss, especially at frequencies above 1 GHz, where the difference is approximately 35 dB compared to the free-space case.

To determine the accuracy of the elliptical model, we compare two cases, namely when the receiver is located on the back of the body and on the shoulder-side. Additionally, in order to improve the elliptical model from the previous section, two muscle tissue arms, composed of cylinders and spheres, were placed alongside the elliptical torso model and are shown superimposed on the actual physical model in Figure 20. The results for the case when the receiver is located on the back of the body and on the shoulder-side are shown in Figures 21 and 22. For the case when the receiver is located on the back of the body the addition of the arms improves the accuracy of the results when compared to the

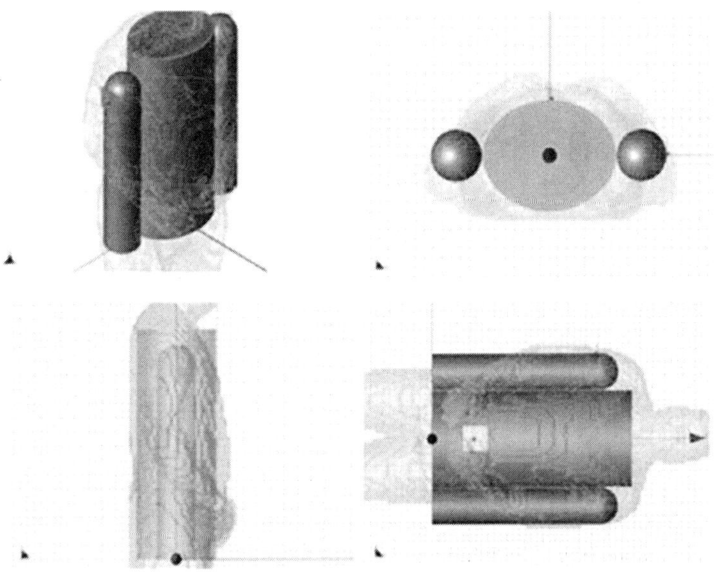

Figure 20. Simplified model superimposed on the human body numerical phantom.

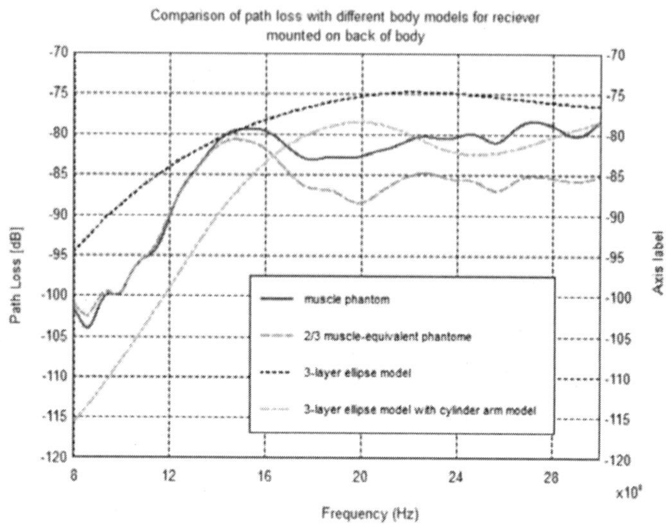

Figure 21. Comparison of the path loss for different phantom models with receiving antenna on the back of the body and transmitting antenna at the waist.

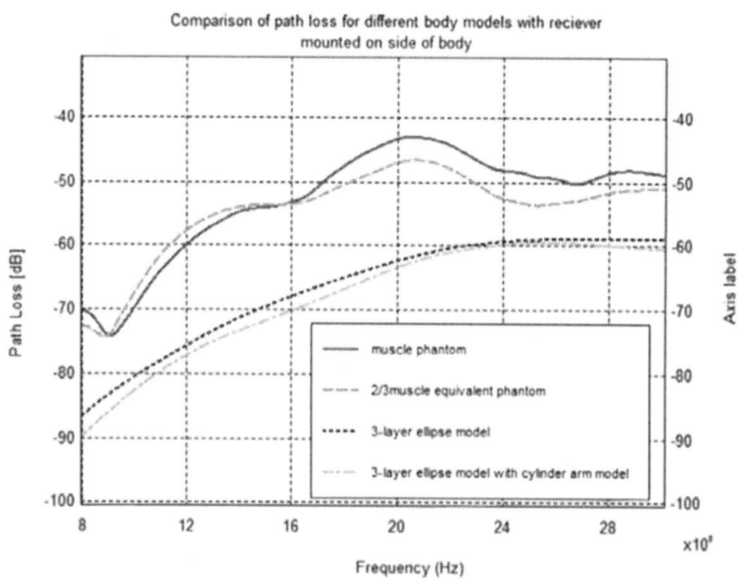

Figure 22. Comparison of the path loss for different phantom models with receiving antenna on the shoulder side of the body and transmitting antenna at the waist.

physical model. This indicates that the arms can have a significant effect on the coupling between antennas mounted on the body. However, for the case when the antenna is placed on the shoulder side of the body, neither the elliptical model nor the ellipse with arms yielded accurate results, and there was approximately a 10 dB discrepancy between the results obtained with these and the physical models. Therefore, we surmise that the addition of other scattering features such as the head, irregular arm position and shape—not exhibited in the simplified models—play a crucial role in accurately capturing the behaviors of the fields around the human body.

6. CONCLUSIONS

Despite the capability of FDTD to model the complex human body in its original form, it is often desirable to seek out good geometrical approximations to simply the computational cost. This has been documented in the literature for both measurement and simulation purposes. However, to-date a quantitative study that demonstrates the accuracy of the results using simplified models has not yet been reported. A detailed characterization of this comparison has been carried out and presented in this paper. It has been shown that the rigorous FDTD phantom model yielded results that are consistent with the measurements found in the literature, and they provide valuable information into potential cosite BAN antenna systems. In addition, we have shown simple models can give reasonably accurate results for some antenna configurations, but they can fail to reproduce the mutual coupling results when the antennas are not in line-of-sight. Therefore, it is highly desirable to carry out an in-depth study of these approximations and to see how well the results based on these approximations compare to the rigorous simulation of the human body.

REFERENCES

1. Alves, T.; Poussot, B.; Laheurte, J.-M. Analytical propagation modeling of BAN channels based on the creeping-wave theory. IEEE Trans. Antennas Propag. **2011**, 59, 1269–1274.

2. Ma, D.; Zhang, W.X. Analytic propagation model for wireless body-area networks. IEEE Trans. Antennas Propag. **2011**, 59, 4749–4756.

3. Dumanli, S.; Railton, C.J. Analysis of coupled tilted slot antennas in FDTD using a novel time domain huygens method with application to body area networks. Trans. Antennas Propag. **2012**, 60, 1987–1994.

4. Eid, A.M.; Wallace, J.W. Accurate modeling of body area network channels using surface-based method of moments. Trans. Antennas Propag. **2011**, 59, 3022–3030.

5. Yu, W.; Mittra, R.; Su, T.; Liu, Y.; Yang, X. Parallel Finite-Difference-Time-Domain Method; Artech House Inc.: Norwood, MA, USA, 2006.

6. Taflove, A.; Hagness, S.C. Computational Electrodynamic—The Finite Difference Time-Domain Method, 3rd ed.; Artech House Inc.: Norwood, MA, USA, 2005; pp. 415–417.

7. Zhao, Y.; Hao, Y.; Alomainy, A.; Parini, C. UWB on-body radio channel modeling using ray theory and subband FDTD method. IEEE Trans. Microwave Theory Tech. **2006**, 54, 1827–1835.

8. Hall, P.S.; Hao, Y. Antennas and Propagation for Body Centric Communications Systems; Artech House: Norwood, MA, USA, 2006; pp. 241–255.

9. Hall, P.S.; Hao, Y.; Nechayev, Y.I.; Alomalny, A.; Constantinou, C.C.; Parini, C.; Kamarudin, M.R.; Salim, T.Z.; Hee, D.T.M.; Dubrovka, R.; et al. Antennas and propagation for on-body communication systems. IEEE Antennas Propag. Mag. **2007**, 49, 41–58.

10. Hall, P.S.; Hao, Y. Wearable Antennas for Body Area Networks. In Microstrip and Printed Antennas: New Trends, Techniques and Applications; John Wiley & Sons: Hoboken, NJ, USA, 2011; pp. 183–210.

11. Gallo, M.; Hall, P.S.; Bai, Q.; Nechayev, Y.I.; Constantinou, C.C.; Bozzetti, M. Simulation and measurement of dynamic on-body communication channels. IEEE Trans. Antennas Propag. Feb. **2011**, 59, 623–630.

12. Alomainy, A.; Hao, Y.; Hu, X.; Parini, C.G.; Hall, P.S. UWB on-body radio propagation and system modelling for wireless body-centric networks. IEE Proc. Commun. **2006**, 153, 107–114.

13. Kamarudin, M.R.; Nechayev, Y.; Hall, P.S. Performance of Antennas in the on Body Environment. Proceedings of the IEEE International Symposium on Antennas and Propagation, Washington, DC, USA, 3– 8 July 2005.

14. Kim, J.; Rahmat-Samii, Y. Implanted antennas inside a human body: Simulations design, and characterization. IEEE Trans. Microw. Theory Tech. **2004**, 52, 1934–1943.

15. Mittra, R.; Bringuier, J.; Wiart, J.; Yoo, K. Modeling of Interaction Between Body-Mounted Antennas. Proceedings of the First European Conference on Antennas and Propagation EuCAP 2006, Nice, France, 6– 10 November 2006.

16. Scanlon, W.G.; Evans, N.E. Body-Surface Mounted Antenna Modelling for Biotelemetry Using FDTD with Homogeneous, Two- and Three-Layer Phantoms. Proceedings of the 10th International Conference on Antennas and Propagation, Edinburgh, Scotland, 14– 17 April 1997.

17. Zasowski, T.; Althaus, F.; Stager, M.; Wittneben, A.; Troster, G. UWAB for Noninvasive Wireless Body Area Networks: Channel Measurements and Results. Proceedings of the IEEE Conference on Ultra Wideband Systems and Technologies, Reston, VA, USA, 16–19 November 2003.

18. Ryckaert, J.; de Doncker, P.; Meys, R.; de Le Hoye, A.; Donnay, S. Channel model for wireless communication around human body. Electron. Lett. **2004**, 40, 543–544.

19. Alomainy, A.; Hao, Y.; Owadally, A.; Parini, C.I.; Nechayev, Y.; Hall, P.S.; Constantinou, C.C. Statistical analysis and performance evaluation for on-body radio propagation with microstrip patch antennas. IEEE Trans. Antennas Propag. **2007**, 55, 245–248.

20. Furse, C.M.; Chen, J.-Y.; Gandhi, O.P. The use of the frequency-dependent finite-difference time-domain method for induced current and SAR calculations for a heterogeneous model of the human body. IEEE Trans. Electromagn. Compat **1994**, 36, 128–133.

21. Gandhi, O.P.; Gao, B.-Q.; Chen, J.-Y. A frequency-dependent finite-difference time-domain formulation for general dispersive media. IEEE Trans. Microw. Theory Tech. **1993**, 41, 658–665.

22. Gabriel, S.; Lau, R.W.; Gabriel, C. The dielectric properties of biological tissues: III parametric models for the dielectric spectrum of tissues. Phys. Med. Biol. **1996**, 41, 2271–2293.

23. Abbasi, Q.H.; Sani, A.; Alomainy, A.; Hao, Y. On-body radio channel characterization and system-level modeling for multiband OFDM ultra-wideband body-centric wireless network. IEEE Trans. Microw. Theory Tech. **2010**, 58, 3485–3492.

24. Khan, M.M.; Abbasi, Q.H.; Alomainy, A.; Hao, Y. Radio Propagation Channel Characterisation Using Ultra Wideband Wireless Tags for Body-Centric Wireless Networks in Indoor Environment. Proceedings of International Workshop on Antenna Technology, Hong Kong, China, 7–9 March 2011.

25. Tuovinen, T.; Yazdandoost, K.Y.; Iinatti, J. Monopole Ultra Wideband antenna for on-body communication in Wireless Body Area Network. Proceedings of the Antennas and

Propagation Conference (LAPC), Loughborough, UK, 14–15 November 2011.

26. Zhang, Y.P.; Li, Q. Performance of UWB impulse radio with planar monopoles over on-human-body propagation channel for wireless body area networks. IEEE Trans. Antennas Propag. **2007**, 55, 2907–2914.

27. Fort, A.; Desset, C.; Ryckaert, J.; de Doncker, P.; van Biesen, L.; Donnay, S. Ultra Wide-band Body Area Channel Model. Proceedings of the IEEE International Conference Communication, Seoul, Korea, 16– 20 May 2005; Volume 4. pp. 2840–1844.

28. Hao, Y.; Alomainy, A.; Song, W.; Parini, C.G.; Hall, P.; Nechayev, Y.; Constantinou, C. Numerical Modeling of On-body Radio Propagation Channel. Presented at the IEEE AP-S International Symposium, Washington, DC, USA, 3– 8 July 2005.

29. Alomainy, A.; Hao, Y.; Parini, C.; Hall, P. Comparison between two different antennas for UWB on-body propagation measurements. IEEE Antennas Wirel. Propag. Lett. **2005**, 4, 31–34.

30. Kamarudin, M.; Nechayev, Y.; Hall, P. Antennas for On-body Communication Systems. Proceedings of the IEEE International Workshop Antenna Technology, Singapore, 7– 9 March 2005; pp. 17–20.

31. Nechayev, Y.I.; Hall, P.S.; Constantinou, C.C.; Hao, Y.; Alomainy, A.; Dubrovka, R.; Parini, C. Antennas and Propagation for On-Body Communication Systems. Proceedings of the 11th International Symposium. on Antenna Technology and Applied Electromagnetics ANTEM, Saint Malo, France, 15–17 June 2005.

32. Nechayev, Y.I.; Hall, P.S.; Constantinou, C.C.; Hao, Y.; Alomainy, A.; Dubrovka, R.; Parini, C. On-Body Path Gain Variations with Changing Body Posture and Antenna Position. Proceedings of the IEEE International Symposium on Antennas and Propagation, Washington, DC, USA, 3– 8 July 2005.

Chapter 11

Interaction of Electromagnetic Radiation with Substance

N. Volobuev Andrey[1]

[1] *Samara State University, Russia*

1. INTRODUCTION

1.1. Distribution of Electromagnetic Field Momentum in dielectrics in stipulation of self-induced transparency

Interaction of quantums of electromagnetic radiation with substance can be investigated both from a wave position, and from a quantum position. From a wave position under action of an electromagnetic wave there are compelled fluctuations of an electronic orbit and nucleus of atoms. The energy of electromagnetic radiation going on oscillation of nucleus passes in heat. Energy of fluctuations of an electronic orbit causes repeated electromagnetic radiation with energy, smaller, than initial radiation.

From a quantum position character of interaction is more various. Interaction without absorption of quantums is possible: resonant absorption, coherent dispersion. The part of quantums is completely absorbed. Quantums can be absorbed without occurrence secondary electrons. Thus all energy of quantums is transferred fonons - to mechanical waves in a crystal lattice, and the impulse is transferred all crystal lattice of substance. At absorption of quantums can arise secondary electrons, for example, at an internal photoeffect. Absorption of quantums with radiation of secondary quantums of smaller energy and frequency is possible, for example, at effect of Compton or at combinational dispersion.

All these processes define as formation of impulses of electromagnetic radiation in substance, and absorption of radiation by substance.

1.2. Coordination of the electromagnetic impulse with the substance

Firstly, consider the one-dimensional task the electric part of electromagnetic field momentum with the dielectric substance, which posses a certain numerical concentration *n* of centrosymmetrical atoms – oscillators. For the certainty of the analysis we suggest the atom to be one-electronic. It is also agreed, that no micro current or free charge are present in the medium. The peculiarities of interaction between magnetic aspect of momentum and the atoms will be considered later.

We accept that there takes place the interaction of quantum of electromagnetic radiation with nuclear electrons, thus quantum are absorbed by the electrons. By gaining the energy of quantum the electrons shift to the advanced power levels. Further, by means of resonate shift of electrons back, appears the quantum radiation forward. The considered medium lacks non-radiating shift of electrons, i.d. the power of quantum is not transfered to the atom.

Thus, the absorption of electromagnetic radiation in the case of its power dissipation in the substance, owing to SIT, is disregarded. There appears the atomic sypraradiation of quantum. Thus, the forefront of momentum passes the power on to the atomic electrons of the medium, forming its back front.

The probabilities of quantum's absorption and radiation by the electrons in the unity of time, with a large quantity of quantum in the impulse, according to Einstein, can be referred to as the approximately identical [6]. For the separate interaction of the with the electron this very probability is the same and is proportional to the cube of the fine-structure constant ~ (1/137)[3] [7]. Consider a random quantity – the number of interactions of quantum with atomic electrons in the momentum. In accordance with the Poisson law of distribution, the probability of that will not be swallowed up any quantum atomic's electrons (will not take place any interaction), at rather low probability of separate interaction, is equal an exponent from the mathematical expectation of a random variable – an average quantity of interactions λ of quantums and electrons in impulse, taken with the minus $p=\exp(-\lambda)$. Therefore, as it will be explained further, it is possible that the intensity of non-absorbed power of impulse by the atomic electrons of the medium in it forefront is determined by the exponential Bouguer law [3] (in German tradition - Beer law)

$$I = I_0 \exp(-\alpha l)\,, \tag{1}$$

where α – index of electromagnetic wave and substance interaction, l – length of interaction layer, I_0 - intensity of incident wave. Thus, the intensity of atomic electron's power recoil into impulse on its back front could be described with the help of the Bouguer law with the negative index of absorption [8].

The index of interaction is $\alpha=\sigma n$, where α – effective section of atom-oscillator interaction with the wave. Hence,

$$\alpha l = \sigma n l = n V_{eff} = n V \frac{V_{eff}}{V} = M \frac{V_{eff}}{V} = MN\,, \tag{2}$$

where V_{eff} – the effective volume of interaction. In defying (1.2) the right part of the formula is multiplied and divided by the geometric volume V, in which there is M of particles interacting with the radiation. The ratio $\frac{V_{eff}}{V} - N$. The ratio of effective volume of interaction to the geometric volume characterizes the medium possibility of electromagnetic radiation's interaction with the atom. Hence, by exponential function in the Bouguer law (1.1) the mathematical expectation of random variable is supposed, which subdues to the Poisson law distribution – average variable of atoms interacting with the electromagnetic radiation in the area of impulse influence $\lambda = NM$.

Taking into account that the wave intensity is $I \sim \begin{pmatrix} E^2 \\ H^2 \end{pmatrix}$ we shall have

$$\begin{pmatrix} |E| \\ |H| \end{pmatrix} = \begin{pmatrix} |E_0| \\ |H_0| \end{pmatrix} \exp\left(-\frac{\alpha}{2} l\right), \tag{3}$$

where $|E0|$, $|H0|$ – the amplitudes of electric and magnetic fields' strength of the impulse on longitudinal coordinate $X = 0$.

In the formula (1.3) and further the upper variables in parentheses are referred to electric field, and lower – to the magnetic field of impulse. By the ratio (1.2) it is possible to find

$$N = -\frac{2}{M} \ln\left|\frac{E}{E_0}\right| = -\frac{2}{M} \ln\left|\frac{H}{H_0}\right| . \tag{4}$$

The formula (1.4) demands some further consideration. If $E<E_0$, that reflects the process of wave absorption by atomic electrons $N>0$ and classical consideration of electromagnetic wave interaction with the atom is quite admissible. The case when $E>E_0$ reflects the process of wave over-radiation. Thus, $N<0$ and variable N can not be considered as the probability of electromagnetic wave interaction with the atom. In this case we speak about the quantum-mechanical character of the process of interaction between the quantum and the bi-level power system of the atom, provided that the power transition's radiation is reversed. Variable N in this case possess the notion of united average of filling by atom $(-1<N<1)$. Due to the use of the average of filling to raise the atom and bend of its magnetic moment in the magnetic field of the impulse, the existence of bi-level quantum system by magnetic quantum numbers. Thus, the variable N provides with the measure of inversion of the system of atom-radiators by the raised atoms [2] as well as the measure of inversion of the magnetic moment of the atom's system by magnetic quantum numbers. If $N=-1$ all the atoms occur in the basic condition [3].

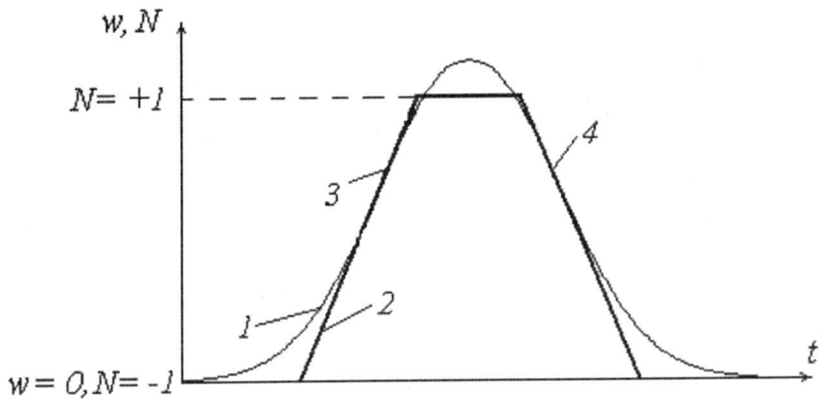

Figure 1. Dependence of volumetric density of energy of electromagnetic radiation impulse *w* (curve 1) and average on atoms of number of filling *N* (curve 2) from time; 3 and 4 - points of an excess of function *w(t)*

We consider the dependence of the average of filling on the time *N(t).* If to accept the proportion of polarization of separate bi-level atom to the intensity of electric field in the impulse, then, in accordance with the Maxwell-Bloch equations, the average by atoms of considered volume, the filling number is proportional to the volumetric density of electromagnetic wave power *N~w* [3]. However such a monotonous dependence between these variables can not remain on the whole extent of the impulse. Firstly, by the high volumetric density of impulse power *w,* typical of SIT, when the central part of impulse power is higher than any variable *w,* there exists energetic saturation of the medium. The average filling number thus *N=1*, all the atoms are raised, fig. 1 (curve 1 - the dependence *w* of time, thicker curve 2 – the considered dependence *N* of time). The violation of proportion *N~w* in the central part of impulse is the basic drawback of frequently used system of Maxwell-Bloch equations for the SIT description.

Secondly, the period of variable *N* relaxation is not less than 1 *ns* [2] that is why the dependence *N(t)* can not repeat high-frequently oscillations on both fronts of the impulse. The dependence *N~w* could characterize the proportion of average filling number and envelope *w* (curve 1) in the impulse. However, in two points of the fold (3 and 4 fig. 1) on the sites of increase and decrease if the envelope *w* the variable ∂2*w*/∂*t*2=0hence, also ∂2*N*/∂*t*2=0. Besides, the dependence *N(t)* has the symmetrical character as at the SIT impulse becomes the conservative system (there is no reverse dispersion and dissipation of power) [2]. Therefore, it could be thoroughly concerned that on the whole extent of impulse, except the points of curve's *N*(*t*) fold, the condition remains

$$\frac{\partial^2 N}{\partial t^2} = 0\,, \tag{5}$$

while the dependence *N(t)* has the character as shown on the fig. 1, curve 2. it could be also highlighted the high generality of formula (1.5), which is possible for any piecewise linear function *N(t)*. Thus, the points of function break are excluded, as the derivates undergo the break.

1.3. Non-linear Schrödinger equation

One-dimensional wave equation for electric and magnetic aspects of electromagnetic field for the considered problem is [2]

$$\frac{\partial^2}{\partial X^2}\begin{pmatrix} E \\ H \end{pmatrix} - \frac{1}{c^2}\begin{pmatrix} \mu \\ \varepsilon \end{pmatrix}\frac{\partial^2}{\partial t^2}\begin{pmatrix} E \\ H \end{pmatrix} = \frac{1}{c^2}\begin{pmatrix} \mu / \varepsilon_0 \\ \varepsilon \end{pmatrix}\frac{\partial^2}{\partial t^2}\begin{pmatrix} P \\ J \end{pmatrix}, \tag{6}$$

where *E≡EY*or*E≡EZ, H≡HY*or*H≡HZ, X* and *t* – accordingly the coordinate alongside of which the impulse and the time are distributed, *P* – polarization of substance, *J* – its magnetization, ε0 and μ0– electrical and magnetic constant, ε – relative static permittivity of substance, μ – relative magnetic permittivity, $c = 1/\sqrt{\varepsilon_0 \mu_0}$ – speed of light in vacuum.

We introduce the transformation of electric field intensity be formula

$$\begin{pmatrix} E(X,t) \\ H(X,t) \end{pmatrix} = \Phi(X,t)\exp(-i\omega_0 t). \tag{7}$$

The function *Φ(X, t)* is less rapidly changing one in time then *E(X,t)* or *H(X,t), ω* $_0$ – aspect of cyclic frequency of high-frequent oscillations of the field.

By substituting (1.7) and (1.6) we get

$$\left[\frac{\partial^2\Phi}{\partial X^2} - \frac{1}{c^2}\begin{pmatrix}\mu \\ \varepsilon\end{pmatrix}\left(\frac{\partial^2\Phi}{\partial t^2} - 2i\omega_0\frac{\partial\Phi}{\partial t} - \omega_0^2\Phi\right)\right]\exp(-i\omega_0 t) = \frac{1}{c^2}\begin{pmatrix}\mu/\varepsilon_0 \\ \varepsilon\end{pmatrix}\frac{\partial^2}{\partial t^2}\begin{pmatrix}P \\ J\end{pmatrix}. \tag{8}$$

We estimate the relative variable of first and second items in the parenthesis of the left side (1.8). for this purpose we would introduce the scales of variables time *t* and *Φ*

$$T = t/t^*, \qquad \Phi_0 = \Phi/\Phi^*,$$

where the asterisk designates dimensionless parameters. For the time scale the duration (period) of impulse *T* should be logically chosen. The scale *Φ* $_0$ is chosen from a condition that dimensionless second derivative $\frac{\partial^2\Phi^*}{\partial t^{*2}}$ and the dimensionless function *Φ** are in the same order. Hence, the the first item in round brackets (1.8) is $\frac{\Phi_0}{T^2}\frac{\partial^2\Phi^*}{\partial t^{*2}}$, and the last one ω20Φ0Φ*. Instead of impulse *T* period we introduce cyclic frequency of impulse $\omega = \frac{2\pi}{T}$. By comparing these items, it is realized, that $\frac{\Phi_0\omega^2}{4\pi^2}\frac{\partial^2\Phi^*}{\partial t^{*2}} << \omega_0^2\Phi_0\Phi^*$ as the cyclic frequency of impulse is far less than infrequences of field's oscillations, especially when $\omega^2 << \omega_0^2$. Similarly, it can be presented that the second item in the round brackets (1.8) is far more that the first one.

Hence, by disregarding the small item in (1.8), we observe

$$\left[\frac{\partial^2\Phi}{\partial X^2} - \frac{1}{c^2}\begin{pmatrix}\mu \\ \varepsilon\end{pmatrix}\left(2i\omega_0\frac{\partial\Phi}{\partial t} + \omega_0^2\Phi\right)\right]\exp(-i\omega_0 t) = \frac{1}{c^2}\begin{pmatrix}\mu/\varepsilon_0 \\ \varepsilon\end{pmatrix}\frac{\partial^2}{\partial t^2}\begin{pmatrix}P \\ J\end{pmatrix}. \tag{9}$$

By accepting vector of polarization *P* or magnetizing *J* to be directly proportional, accordingly, to the electric and magnetic fields strength, we could derive the wave equation from (1.6), which is possible to any form of the wave. However, there exists a physical mechanism, which restrict the wave form. This mechanism is connected with the way of over-radiating of electromagnetic impulse with the atomic electrons. This process is precisely considered further.

We consider the strength of electric and magnetic fields of impulse as

$$\begin{pmatrix} E(X,t) \\ H(X,t) \end{pmatrix} = \begin{pmatrix} |E(X,t)| \\ |H(X,t)| \end{pmatrix} \exp[i(rX - \delta t)], \tag{10}$$

where *r* and δ – are constants, */E (X,t)/* and */H(X,t)/* are the modules of functions *E(X,t)* and *H(X,t).*

Formulas (1.4) and (1.5) reflect the offered physical model of electric and magnetic field of impulse interaction with atoms in SIT.

Hence, taking into account (1.4) and (1.5) there is

$$\frac{\partial^2 \ln\left|\frac{E}{E_0}\right|}{\partial t^2} = \frac{\partial^2 \ln\left|\frac{H}{H_0}\right|}{\partial t^2} = 0 . \tag{11}$$

By transforming (1.11) we have

$$\frac{\partial^2 |E|}{\partial t^2} = \left(\frac{\partial \ln|E|}{\partial t}\right)^2 |E| . \tag{12}$$

The similar ratio can be also referred to the function $|H|$. These ratios should not be regarded as the equations to define the module of electric and magnetic aspect of impulse. It is the approximate expression of the second derivative $\frac{\partial^2 |E|}{\partial t^2}$ or $\frac{\partial^2 |H|}{\partial t^2}$ for the considered physical model and reflects several non-linear effects of interaction between electromagnetic radiation and substance. The approximate ratio (1.12) defines the connection of medium polarization *P* with the strength of impulse electric field (similarly to the magnetization J with the magnetic field strength), that would be considered further. The electromagnetic field impulse strengths should be estimated from the equation (1.6) taking into account the ratio (1.12).

In accordance with (1.10),

$\begin{pmatrix} |E(X,t)| \\ |H(X,t)| \end{pmatrix} = \begin{pmatrix} E(X,t) \\ H(X,t) \end{pmatrix} \exp[-i(rX - \delta t)]$, , hence, from (1.12) we estimate equation for the electromagnetic field impulse

$$\frac{\partial^2 E}{\partial t^2} = -2i\delta\frac{\partial E}{\partial t} + \left[\delta^2 + \left(\frac{\partial \ln\left|\frac{E}{E_0}\right|}{\partial t}\right)^2\right] E .$$

(13)

The same ratio exists for the magnetic field also. Passing over to (1.13) to the function $\Phi(X,t)$ by formula (1.7) and by concerning $P=\varepsilon 0\chi E$, where χ – relative dielectric permittivity of substance, we have

$$\frac{\partial^2 P}{\partial t^2} = \left[-2i\delta\varepsilon_0\chi\frac{\partial \Phi}{\partial t} - 2\delta\varepsilon_0\chi\omega_0\Phi + \varepsilon_0\chi\left[\delta^2 + \left(\frac{\partial \ln\left|\frac{\Phi}{\Phi_0}\right|}{\partial t}\right)^2\right]\Phi\right]\exp(-i\omega_0 t),$$

(14)

For the variable $\frac{\partial^2 J}{\partial t^2}$ by using $J = \chi H$, , where χ – relative magnetic permittivity of substance, we get the ratio, similar to (1.14), except that the right part lacks ε 0.The variables $\begin{pmatrix} E_0 \\ H_0 \end{pmatrix} = \Phi_0 \exp(-i\omega_0 t)$. By comparing (1.7) and (1.10) we state $\begin{pmatrix} |E| \\ |H| \end{pmatrix} = |\Phi|$, $\begin{pmatrix} |E_0| \\ |H_0| \end{pmatrix} = |\Phi_0| = const$.

By substituting (1.14) into (1.9)

$$2i(\omega_0 + \chi\delta)\frac{\partial \Phi}{\partial t} + c^2\begin{pmatrix} 1/\mu \\ 1/\varepsilon \end{pmatrix}\frac{\partial^2 \Phi}{\partial X^2} + \left(\omega_0^2 + 2\chi\delta\omega_0 - \chi\delta^2\right)\Phi = \chi\left(\frac{\partial \ln\left|\frac{\Phi}{\Phi_0}\right|}{\partial t}\right)^2\Phi .$$

(15)

In the equation (1.15) the variable χ is meaningful to dielectric permittivity for electric and magnetic permittivity for the magnetic aspects of electromagnetic field.

The non-linear Schrödinger equation with complicated type of linearity is received. We introduce the signs: $\alpha = \omega_0 + \chi\delta$, $\gamma = \omega_0^2 + 2\chi\delta\omega_0 - \chi\delta^2 = \alpha^2 - \begin{pmatrix}\varepsilon \\ \mu\end{pmatrix}\chi\delta^2$, where $\begin{pmatrix}\varepsilon \\ \mu\end{pmatrix} = 1 + \chi$ – – relative permittivities of the substance. Hence, the equation (1.15) will be

$$2i\alpha\frac{\partial\Phi}{\partial t} + c^2\begin{pmatrix}1/\mu \\ 1/\varepsilon\end{pmatrix}\frac{\partial^2\Phi}{\partial X^2} + \gamma\Phi = \chi\left(\frac{\partial \ln\left|\frac{\Phi}{\Phi_0}\right|}{\partial t}\right)^2\Phi. \tag{16}$$

We shall find the solution to the non-linear Schrödinger equation (1.16) as in [9]

$$\Phi = \Phi_0 f(kX - \omega t)\exp\left[i\left(rX - \delta^* t\right)\right], \tag{17}$$

where the type of the function $f(kX-\omega t)$ is still unknown. The variables k, ω and δ^* – constants. By marking $\zeta = kX - \omega t$, and substituting (1.17) in (1.16) and concerning $|\Phi| = \Phi 0 f(\zeta)$ we get

$$c^2\begin{pmatrix}1/\mu \\ 1/\varepsilon\end{pmatrix}k^2\frac{d^2 f}{d\zeta^2} + 2i\frac{df}{d\zeta}\left(krc^2\begin{pmatrix}1/\mu \\ 1/\varepsilon\end{pmatrix} - \alpha\omega\right) + f\left(\gamma + 2\alpha\delta^* - r^2c^2\begin{pmatrix}1/\mu \\ 1/\varepsilon\end{pmatrix}\right) = \chi\omega^2\left(\frac{d\ln f}{d\zeta}\right)^2 f \tag{18}$$

If to permit that $krc^2 = \alpha\begin{pmatrix}\mu \\ \varepsilon\end{pmatrix}\omega$ as there should not be any imaginary items in (1.18), this equation is transformed to

$$c^2\begin{pmatrix}1/\mu \\ 1/\varepsilon\end{pmatrix}k^2\frac{d^2 f}{d\zeta^2} + f\left(\gamma + 2\alpha\delta^* - r^2c^2\begin{pmatrix}1/\mu \\ 1/\varepsilon\end{pmatrix}\right) = \chi\omega^2\left(\frac{d\ln f}{d\zeta}\right)^2 f. \tag{19}$$

We consider the solution of the equation (1.19) by

$$f = C_1\exp\left[\frac{C_2\zeta^2}{4}\right], \tag{20}$$

where C_1 and C_2 – constants. By substituting (1.20) into (1.19) we get that the constant C_1 could be the arbitrary variable, $\chi\begin{pmatrix}\mu\\ \varepsilon\end{pmatrix}\omega^2 = k^2c^2$.

The constant C_2 could not depend upon the parameters of equation. It is accepted that C_2 =*-1*. Then the frequency and the wave number in (1.17), accordingly, are

$$\delta^* = \frac{c^2\begin{pmatrix}1/\mu\\ 1/\varepsilon\end{pmatrix}(r^2+k^2/2)-\gamma}{2\alpha}; \qquad r = \frac{\alpha\begin{pmatrix}\mu\\ \varepsilon\end{pmatrix}\omega}{kc^2}. \tag{21}$$

The formulas (1.21) associate the frequency and the wave number of oscillations of function *Φ(X,t)* with the parameters of substance and electromagnetic field impulse.

The most simple ratios between the parameters are gained, when $\delta = \omega 0$. In this case $\alpha = \begin{pmatrix}\varepsilon\\ \mu\end{pmatrix}\delta$, $\gamma = \begin{pmatrix}\varepsilon\\ \mu\end{pmatrix}\delta^2$. From the equations in (1.21), and concerning $\chi\begin{pmatrix}\mu\\ \varepsilon\end{pmatrix}\omega^2 = k^2c^2$ there is

$$r = \frac{\alpha\begin{pmatrix}\mu\\ \varepsilon\end{pmatrix}\omega}{kc^2} = \frac{\alpha k}{\chi\omega} = \frac{\begin{pmatrix}\varepsilon\\ \mu\end{pmatrix}\delta k}{\chi\omega}, \quad \delta^* = \frac{\alpha}{2\chi} + \frac{\chi\omega^2}{4\alpha} - \frac{\gamma}{2\alpha} - \frac{2\begin{pmatrix}\varepsilon\\ \mu\end{pmatrix}\delta^2 + \chi^2\omega^2}{4\chi\begin{pmatrix}\varepsilon\\ \mu\end{pmatrix}\delta} \tag{23}$$

By concerning that $2\begin{pmatrix}\varepsilon\\ \mu\end{pmatrix}\delta^2 >> \chi^2\omega^2$, we have $\delta^* \approx \frac{\delta}{2\chi}$. This inequality is true, as for the rarefied gas (*n* < 10^{18} *atoms/cm* ^{3}) $\begin{pmatrix}\varepsilon\\ \mu\end{pmatrix} >> \chi$ and the frequency of wave filling of impulse δ is far more than frequency of impulse envelope ω.

Taking into account (1.10), (1.20) and the $\begin{pmatrix}|E|\\ |H|\end{pmatrix} = |\Phi|$, we can find the laws of electromagnetic field strengths shifting by

$$\begin{pmatrix} E \\ H \end{pmatrix} = \begin{pmatrix} E_0 \\ H_0 \end{pmatrix} \exp\left[-\frac{(kX-\omega t)^2}{4} \right] \exp\left[i(rX-\delta t) \right].$$

(24)

It should be stressed, that though, the ratios for the electric aspect of impulse in [1] and (1.23) are similar to each other and feature the same phases of oscillations, that is possible on some distance from the over-radiating atom, the non-linear Schrödinger equations are differ in type of non-linearity. The reason of this lies in the fact that in [1] the impulse was considered with regard to low intensity, the one that does not lead to the energetic saturation of medium, in which it is disseminated.

For the estimation, like in [1] we have $k = \frac{\omega}{c} = 2{,}1 \cdot 10^4 m^{-1}$, $r = \frac{\delta}{c} = 2{,}1 \cdot 10^5 m^{-1}$, ω=6,28·10[12] s^{-1}, δ= 6,28·10[13] s^{-1}.

For instance, the result of strength estimation of the electric filed impulse by the coordinate X, calculated with the MathCAD system by formula (1.23), is shown in fig.2.

Taking in to account the reciprocal orthogonality of planes of vectors' envelopes of electric and magnetic fields impulse, we could gain the type of electromagnetic soliton, fig. 3.

Figure 4 shows the envelopes of electric field impulse in the SIT, based on formula (1.23), curve 1, and by formula (1.24), being the consequence of Maxwell-Bloch theory, curve 2. the impulse envelope of electric field strength in this theory is expressed as the first derivative of the Sin-Gordon equation solving and is

$$E = \frac{E_0}{\mathrm{ch}(kX-\omega t)}.$$

(24)

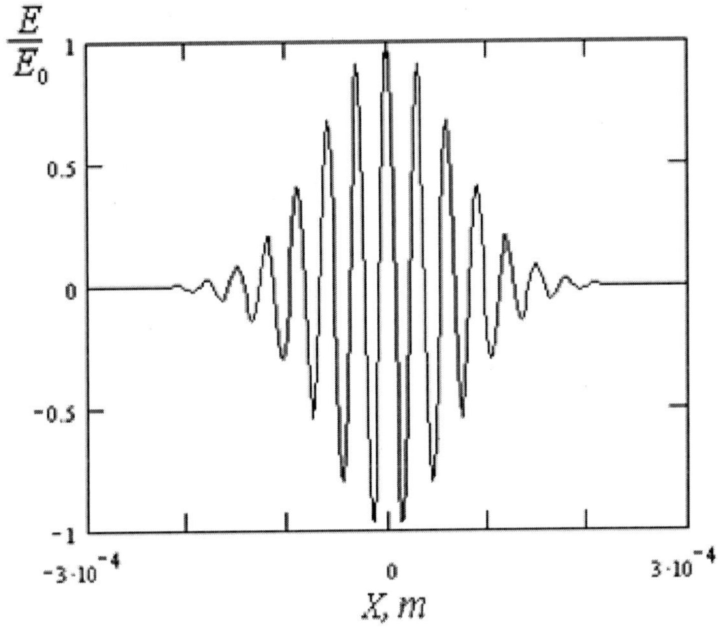

Figure 2. Calculation of the electric component of electromagnetic radiation impulse in dielectric.

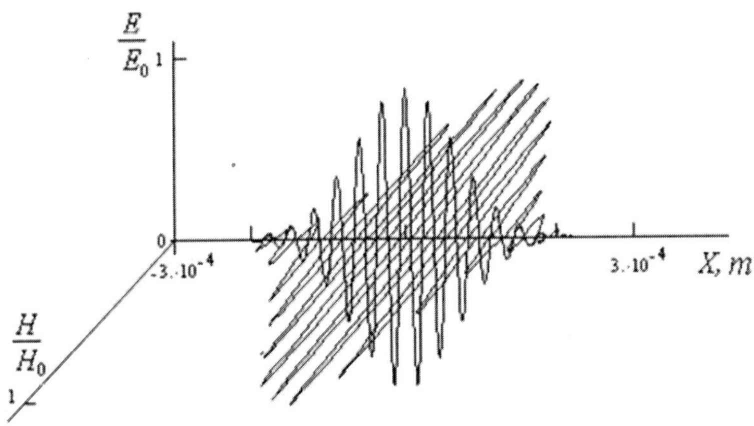

Figure 3. Intensity of electric and magnetic fields electromagnetic solitone in dielectric in conditions of the self-induced transparency.

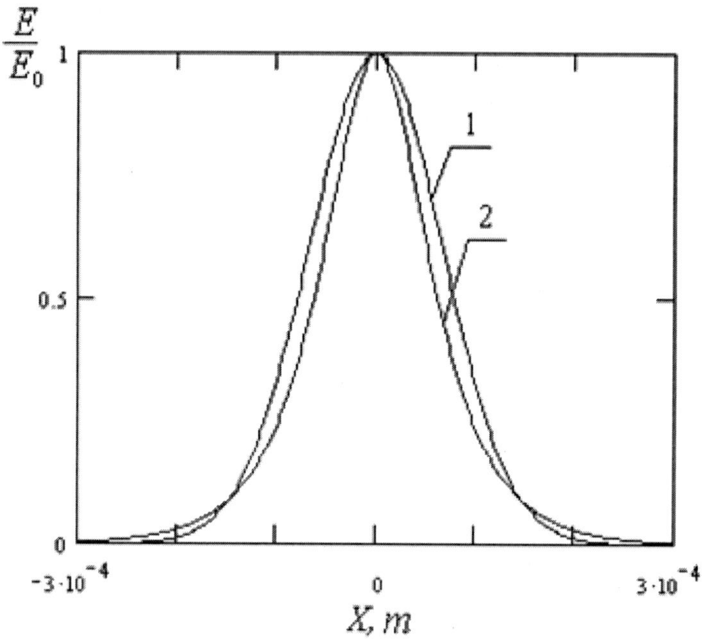

Figure 4. Comparison bending around of the electromagnetic field impulse, received on the basis of the offered theory, a curve 1, and the equations the Maxwell - Bloch, curve 2

Evidently, the first derivative of Sin-Gordon equation solving is similar to the soliton envelope in the non-linear Schrödinger equation with cube non-linearity solving (27). Curves 1and 2 in fig. 4 are designed for the same parameters as the function in fig. 2. We can infer from fig. 4 that impulse, referred to formula (1.23), curve 1, is broader in its central part, but asymptotically shorter than impulse, inferred by the Maxwell-Bloch theory, curve 2. Evidently, its is bound with the energetic permittivity of medium in the central part of impulse.

2. ANGULAR DISTRIBUTION OF PHOTOELECTRONS DURING IRRADIATION OF METAL SURFACE BY ELECTROMAGNETIC WAVES

There is the problem of achieving of the maximum photoelectric flow during irradiation of the metal by flow of electromagnetic waves while designing of photoelectrons. The depth of radiation penetration into metal during irradiation of its surface is defined by the Bouguer low [10]:

$$I = I_0 \exp\left(-\frac{4\pi}{\lambda} n\chi z\right),$$

where I_0 – is the intensity of the incident wave, I – is the intensity on z-coordinate, directioned depthward the metal, λ – is the wavelength of radiation, $n\chi$ – is the product of refractive index by extinction coefficient.

Let's estimate the thickness of the metal at which intensity of light decreases in e = 2,718 times:

$$z = \frac{\lambda}{4\pi n\chi}$$

Average wavelength of a visible light for gold λ=550 nm, $n\chi$=2,83, therefore z = 15,5 nm. Considering [11] that lattice constant for gold a = 0,408 nm, it is possible to deduce that electromagnetic radiation penetrates into the metal on 40 atomic layers.

Therefore radiation interaction occurs basically of the top layers of atoms and angular distribution of electron escape from separate atoms, i.e. during the inner photoemissive effect, it will appreciably have an impact on distribution of electron escape from the metal surface.

As a result it is interesting to consider angular distribution of photoelectrons during the inner photoemissive effect.

2.1. Nonrelativistic case

Although Einstein has explained the photoeffect nature in the early 20th century, various aspects of this phenomenon draw attention, till nowadays for example, the role of tunnel effect is investigated during the photoeffect [12].

In the description of angular distribution of the photoelectrons which are beaten out by photons from atoms, there are also considerable disagreements. For example it is possible to deduce that the departure of photoelectrons forward of movement of the photon and back in approach of the main order during the unitary photoeffect is absent, using the computational method of Feynman diagrams [13]. I is marked that photoelectrons don't take off in the direction of distribution of quantum [14]. This conclusion is made on the basis of positions which in the simplified variant are represented by the following.

The momentum of the taken off electron is defined basically by action produced by the electric vector of quantum of light on electron. If electron takes off in the direction of an electric vector of quantum it gets the momentum. On a plane set at an angle to a plane of polarization of quantum of light, (fig. 5) electron momentum value will be $p1m{=}pe\cos\phi$.

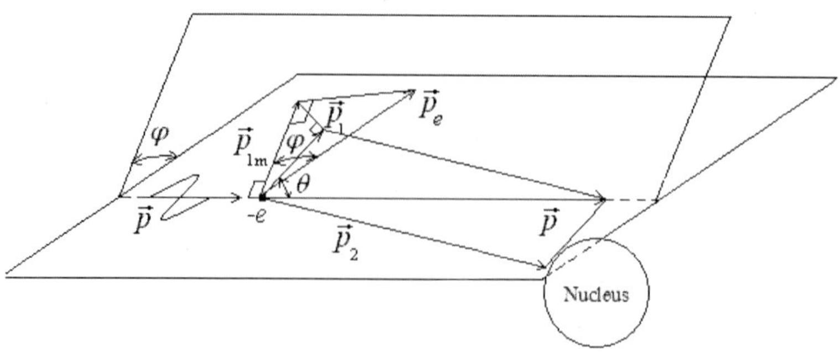

Figure 5. Direction of vectors of particle momentum at the inner photoemissive effect

Besides, if the electron momentum is set at an angle θ to the direction of quantum of light its value will be:

$$p_1 = p_e \cos\phi \sin\theta \tag{2.1}$$

Therefore, photoelectron energy is equal to:

$$E_1 = \frac{p_1^2}{2m_1} = \frac{p_e^2 \cos^2\phi \sin^2\theta}{2m_1}, \tag{2.2}$$

where m_1 – is the electronic mass.

If ϑ=0 then photoelectron energy $E1$=0 . Photoelectrons take off readies its maximum in the direction of a light vector or a polarization vector, i.e. an electric field vector of quantum of light. The same dependence is offered in the work [7]. The formula (2.2) has the simplified nature in comparison with [7, 14], but convey correctly the basic dependence of distribution energy of a photoelectrons escape from the corners ϕ and θ.

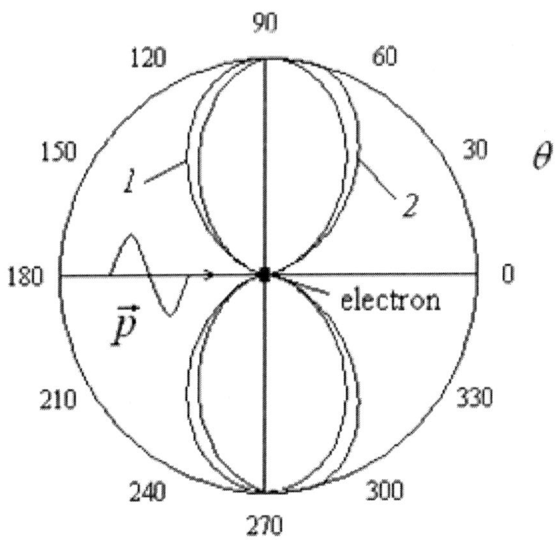

Figure 6. Angular distribution of photoelectrons during interaction of orbital electron with the electromagnetic wave

The lack of dependence (2.2) is that at its conclusion the law of conservation of momentum, wasn't used and therefore there is no electron movement to the direction $\vartheta=0$. Usage of the momentum conservation equation in [7, 14] can't be considered satisfactory since in the analysis made by the authors it has an auxiliary character. At the heart of the analysis [7, 14] is the passage of electron from a discrete energy spectrum to a condition of a continuous spectrum under the influence of harmonious indignation, i.e. the matrix element of the perturbation operator is harmonious function of time. In other words, the emphasis is on the wave nature of the quantum cooperating with electron. Angular distribution of electron energy in the relative units, made according the formula (2.2) is shown on fig. 6, a curve 1. Let's illustrate the correction to the formula (2.2) connected with presence of photon momentum, following [15].

Fig.7 demonstrates change of photoelectron momentum in the presence of a photon momentum. The conclusion made on the basis of the is $\vartheta=\vartheta'+\delta$. Let's find $\sin\vartheta=\sin\vartheta'\cos\delta+\sin\delta\cos\vartheta'$. Considering that δ is too small we find $\sin\theta=\sin\theta'\left(1+\frac{\sin\delta\cos\theta'}{\sin\theta'}\right)=\sin\theta'\left(1+\frac{p}{p_e}\cos\theta'\right)$. The law of sines for a triangle on fig. 7 is used.

Further consideration $\beta'=\frac{p}{p_e}=\frac{h\nu}{mVc}=\frac{1}{2}\beta+\frac{W}{mVc}$,

where $\beta=\frac{V}{c}$ – is the relation of photoelectron speed to a speed of light in vacuum, W – is the work function of electrons from atom, we have $\sin\vartheta=\sin\vartheta'(1+\beta'\cos\vartheta')$. Taking for granted that β′ is small we will transform (2.2) into $E_1=\frac{p_e^2\cos^2\phi\sin^2\theta'}{2m_1}\left(1+2\beta'\cos\theta'\right)$. Angular distribution of electron energy for $\beta'=0{,}15$, made according to the (2.2) taking into account the correction is shown on fig. 7, a curve 2.

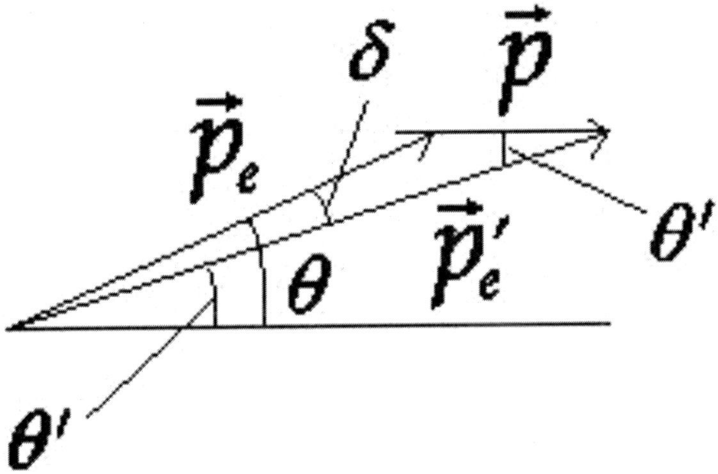

Figure 7. The account of the momentum of quantum $\vec{p}$ using wave approach of electron Interaction with the electromagnetic wave

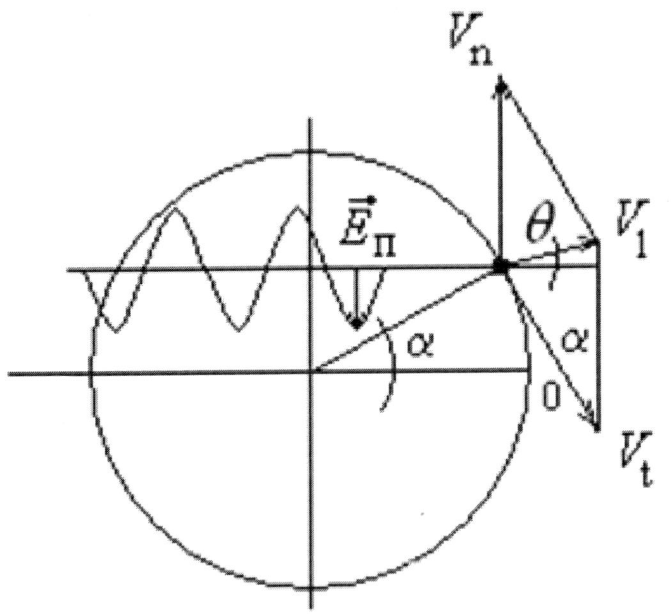

Figure 8. Attitude of components velocity of orbital electron during its interaction with the electromagnetic wave

Thus scattering indicatrix of photoelectrons has received some slope forward, but to the direction of quantum momentum, i.e. at ϑ=0 electrons don't take off as before. The formula (2.2) is accounted as a basis of the wave nature of light. For the proof of this position we will consider interaction of an electromagnetic wave with orbital electron. The description of orbital movement electron is done on the basis of Bohr semiclassical theory since interacting process of electron with an electromagnetic wave is investigated from the positions of classical physics, fig. 8.

By the sine law from a triangle of speeds we find:

$$\sin\alpha = \frac{V_1}{V_t}\cos\theta \ , \tag{2.3}$$

where V_t – is the speed of electron movement round the nucleus, V_1 – is the total speed of electron considering the influence on it of an electromagnetic wave.

By the law of cosines we have:

$$V_1^2 = V_n^2 + V_t^2 - 2V_n V_t \cos\alpha = V_n^2 + V_t^2 - 2V_n V_t \sqrt{1-\left(\frac{V_1}{V_t}\cos\theta\right)^2} \ , \tag{2.4}$$

where V_n – is the component of the general speed of electron movement after its detachment from a nucleus which arises under the influence of dielectric field intensity $\vec{E}$ in the electromagnetic wave.

Solving (2.4) rather V_1, we find:

$$\left(V_1^2\right) = \left(V_n^2 + V_t^2 - 2V_n^2\cos^2\theta\right) + \sqrt{\left(V_n^2 + V_t^2 - 2V_n^2\cos^2\theta\right)^2 - \left(V_n^2 - V_t^2\right)^2} \ . \tag{2.5}$$

The condition of detachment electron from atom at any position of electron $Vn \geq Vt$ In case of equality of speeds

$Vn=Vt$

we have:

$$V_1 = 2V_{II} \sin\theta \, . \tag{2.6}$$

Distribution of speeds (2.6) corresponds to (2.2) and fig. 6, a curve 1. Thus, the parity (2.6) arises if to consider only the wave nature of the electromagnetic wave cooperating with orbital electron.

In [6] distribution of an angle of the electron escape is investigated only for a relativistic case. It is thus received that electrons are emanated mainly to a direction of photon distribution. However the done conclusion is also actually based on the formula (2.1). Therefore the drawback of the conclusion [6] is in absence in definitive formulas of angular distribution of electrons of nuclear mass m_2. And after all the nuclear mass defines a share of the photon momentum which can incur a nuclear.

Let's consider the phenomenon of the inner photoemissive effect from positions of corpuscular representation of quantum of light, fig. 5. The quantum of light by momentum $\vec{p}$ and energy E beats out electron from atom, making A a getting out. Thus both laws of conservation of energy should be observed:

$$E = A + E_1 + E_2 \, , \tag{2.7}$$

Where E_1 – is the kinetic energy of taken off electron, E_2 – is the kinetic energy of nucleus as well as the law of conservation of momentum:

$$\vec{p} = \vec{p}_1 + \vec{p}_2 \, , \tag{2.8}$$

Where $\vec{p}\,1$ - is the momentum of taken off electron, $\vec{p}\,2$– is the momentum transferred to a nucleus. The formula (2.7) differs from Einstein's standard formula $E=A+E1$. The point is that Einstein's formula means the absence of angular distribution of photoelectrons speed. Really, if energy of photon E is set and work function A for the given chemical element is determined certain speed of the electron escape from atom is thereby set. It means that speeds of electrons, taking off to every possible directions are identical, and the problem of finding out their angular distribution is becoming incorrect. The value of the momentum transferred to a nucleus can be found using the formula, following (2.8):

$$p_2^2 = p^2 + p_1^2 - 2pp_1\cos\theta . \tag{2.9}$$

The system of equations (2.7) and (2.9) to obtain a combined solution and the equation (2.9) are convenient to express through energy. Taking into account *E=pc* , where c – is the speed of light in vacuum, $p_1^2 = 2m_1E_1$ and $p_2^2 = 2m_2E_2$, we find:

$$2m_2E_2 = \left(\frac{E}{c}\right)^2 + 2m_1E_1 - 2\frac{E}{c}\sqrt{2m_1E_1}\cos\theta , \tag{2.10}$$

where m_1 – is the electronic mass, m_2 – is the nuclear mass.

Substituting in (2.10) kinetic energy of nuclear E_2 by (2.7), we have:

$$E - A - E_1 = \frac{1}{2m_2}\left(\frac{E}{c}\right)^2 + \frac{m_1}{m_2}E_1 - \frac{E}{m_2c}\sqrt{2m_1E_1}\cos\theta . \tag{2.11}$$

Let us introduce the following notation $G = \sqrt{E_1}$, $\alpha = \frac{E}{m_2c}\sqrt{2m_1}$, $\sigma = 1 + \frac{m_1}{m_2}$, $\gamma = \frac{1}{2m_2}\left(\frac{E}{c}\right)^2 + A - E$. Then the equation (2.11) will be transformed into:

$$\beta G^2 - \alpha G\cos\theta + \gamma = 0 . \tag{2.12}$$

Solving quadratic equation (2.12) provided $\sigma \approx 1$ (electronic mass is much less that nuclear mass), we find:

$$G_{1,2} = \frac{\alpha}{2}\cos\theta \pm \sqrt{\frac{\alpha^2}{4}\cos^2\theta - \gamma} . \tag{2.13}$$

Substituting in (2.13) accepted notation we have:

$$G_{1,2} = \sqrt{\frac{m_1}{2}}\frac{E}{m_2c}\cos\theta \pm \sqrt{\frac{E^2}{2m_2c^2}\left(\frac{m_1}{m_2}\cos^2\theta - 1\right) + E - A} . \tag{2.14}$$

Considering that $G=\sqrt{E_1}=V_1\sqrt{\frac{m_1}{2}}$, where V_1 – speed of photoelectrons provided $\frac{m_1}{m_2}\cos^2\theta \ll 1$, we find:

$$V_1=\frac{E}{m_2c}\cos\theta \pm \sqrt{\left(\frac{m_1}{m_2}\cos^2\theta-1\right)\frac{m_2}{m_1}\left(\frac{E}{m_2c}\right)^2+\frac{2(E-A)}{m_1}}. \tag{2.15}$$

Provided that nuclear mass is aiming to infinity $m2\to\infty$ the formula (2.15) is transformed into Einstein's standard law for the photoeffect. Besides, this, as if it has been specified earlier, angular distribution of speed of photoelectrons disappears.

The condition $m2\to\infty$ is fair in outer photoemissive effect when the photon momentum is transferred to the whole metal through single atoms. Therefore for an outer photoemissive effect, i.e. for interaction of the solid and the photon, Einstein's formula $E=A+E1$ is applicable absolutely.

For the inner photoemissive effect in the formula (2.15) it is necessary to use effective nuclear mass *m2effm2* , considering attractive powers between atoms in substance.

Transforming the formula (2.15), we get:

$$V_1=\frac{E}{m_2c}\left(\cos\theta \pm \sqrt{\cos^2\theta+\frac{m_2}{m_1}\left(2(E-A)\frac{m_2c^2}{E^2}-1\right)}\right). \tag{2.16}$$

Let us nominate $\Delta=E-A$. Distribution of photoelectrons will arise at $\Delta \geq \frac{E^2}{2m_2c^2}$. In the right part of the received inequality there is a very small value, therefore distribution of photoelectrons will arise practically at *EA*.

Let us nominate $\Delta=\eta\frac{E^2}{2m_2c^2}$, where $\eta\geq 1$ characterizes the value of exceedance of photon energy over work function in relative units. Thus the formula (2.16) takes the form:

$$V_1 = \frac{E}{m_2 c}\left(\cos\theta + \sqrt{\cos^2\theta + \frac{m_2}{m_1}(\eta - 1)}\right). \tag{2.17}$$

The analysis of the formula (2.17) shows that the root must to taking a plus since otherwise electron scattering basically goes aside, contrary to the direction of a falling photon. Angular distribution of the electron escape during the inner photoemissive effect in the relative units $\frac{V_1}{E/m_2c}$ is shown on fig. 9, made according to the formula (2.17) with several values η for copper.

№	η
1	1
2	1,00001
3	1,0001
4	1,01

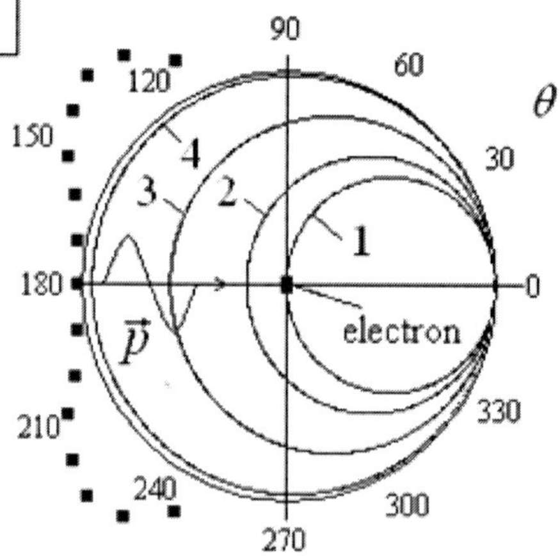

Figure 9. Angular distribution of photoelectrons during the inner photoemissive effect depending on parameter η during interaction of orbital electrons with light quantum. The results of experiments [16] are shown by black small squares

The figure makes it evident that speeds of photoelectrons become almost identical in all directions already at $\eta \geq 1{,}01$. Then Einstein's formula *E*=*A*+*E*1becomes fair and for the inner photoemissive effect.

Considering that, for example, for copper the relation $\frac{E^2}{2m_2c^2} \Big/ E \approx 4{,}2 \cdot 10^{-11}$ equivalent as far as the order of value is concerned $\frac{E^2}{2m_2c^2} \Big/ \Delta = \frac{1}{\eta}$ in the field of red photoelectric threshold (λ_r = 250 nm), it is possible to draw the conclusion that the evident difference of distribution of photoelectrons speeds from spherical, i.e. actually formula is violated $E=A+E1$, can be observed only in very short wave part of spectrum γ-radiations.

The observed data of angular distribution of the photoelectrons which have been beaten out from a monolayer of atoms of copper by covering the nickel surface are shown in fig. 9 by black small squares [16]. The wavelength of quanta allowed observing the photoeffect with 2p-atom shell of copper, but the photoeffect on nickel thus was absent. Experimental distribution of photoelectrons contradicts calculated distribution in fig. 6. Moreover, in distinction in fig. 6, small maxima of indicatrix of the distributions directed to an opposite direction of flight of light quanta at an angle of approximately 45º to the direction of light flux are observed. In [16] these maxima are explained by focusing properties of all population of atoms of the surface. The amplitude of maxima ascends with the increase of quantity of the monolayers of copper atoms on nickel.

Thus, angular distribution of photoelectrons will be absolutely various depending on whether what properties, wave or corpuscular are reveal by the light quantum in interaction with orbital electron. Only experiment can give the answer to the question what distribution it is true, fig. 6 or fig. 9. However existence of electron flux from an illuminated surface at normal light incidence [16], in the direction opposite to intensity of light, shows at the prevalence of corpuscular properties of light in its interaction with atoms.

2.2. Relativistic case

Dealing with relativistic case of the inner photoemissive effect, the law of conservation of energy needs to be written down as:

$$E = A + E_{\text{k}} + E_2 \,, \tag{2.18}$$

Where *E*k– is the kinetic energy of photoelectron.

The law of conservation of momentum remains in the form (2.9). Using relativistic relation between the energy and the momentum for electron:

$$E_1^2 = p_1^2 c^2 + m_1^2 c^4 \,, \tag{2.19}$$

where E_1 – is the total energy of electron, m_1 – is the electron rest mass, we will express the momentum of electron from (2.19)and we will substitute in (2.9). For convenience of the further transformations we will write down (2.19) into:

$$p_1^2 = \frac{E_1^2 - m_1^2 c^4}{c^2} = \frac{\left(E_1 + m_1 c^2\right) E_{\text{k}}}{c^2} \,. \tag{2.20}$$

Formulating (2.20) the relation has been used:

$$E_{\text{k}} = E_1 - m_1 c^2 \,. \tag{2.21}$$

The equation (2.9) will be transformed into:

$$2 m_2 c^2 E_2 = E^2 + \left(E_1 + m_1 c^2\right) E_{\text{k}} - 2E \sqrt{\left(E_1 + m_1 c^2\right) E_{\text{k}}} \cos\theta \,. \tag{2.22}$$

Because of that the nucleus that has a big mass and a relatively low speed after interaction with the photon, expression for relation of the momentum of the nucleus with its kinetic energy E_2 is used in the nonrelativistic form.

Substituting value E_2 in (2.22) from the equation (2.18), we get:

$$2 m_2 c^2 \left(E - A - E_{\text{k}}\right) = E^2 + \left(E_1 + m_1 c^2\right) E_{\text{k}} - 2E \sqrt{\left(E_1 + m_1 c^2\right) E_{\text{k}}} \cos\theta \,. \tag{2.23}$$

Let us nominate:

$$\alpha = \frac{E\sqrt{E_1 + m_1c^2}}{m_2c^2}; \qquad G = \sqrt{E_k}\ . \tag{2.24}$$

As a result (2.23) will be transformed into:

$$\delta G^2 - \alpha G\cos\theta + \gamma = 0\ . \tag{2.25}$$

The notation $\gamma = \frac{1}{2m_2}\left(\frac{E}{c}\right)^2 + A - E = \frac{1}{2m_2}\left(\frac{E}{c}\right)^2 - \Delta = \frac{1}{2m_2}\left(\frac{E}{c}\right)^2(1-\eta)$ corresponds to item 1 section.

The value

$$\delta = 1 + \frac{m_2c^2}{2}\left(\frac{\alpha}{E}\right)^2 = 1 + \frac{m_1}{2m_2} + \frac{E_1}{2m_2c^2} = 1 + \frac{m_1}{2m_2} + \frac{E_k - m_1c^2}{2m_2c^2} = 1 + \frac{E_k}{2m_2c^2} \approx 1\ .$$

It is thus accounted for that $E_k << 2m_2c^2$.

Solving the equation (2.25), we get:

$$G_{1,2} = \frac{\alpha}{2}\cos\theta \pm \sqrt{\frac{\alpha^2}{4}\cos^2\theta - \gamma}\ . \tag{2.26}$$

Substituting notations, we find:

$$E_k = \left(\frac{\alpha}{2}\right)^2\left(\cos\theta + \sqrt{\cos^2\theta + \frac{2}{\alpha^2}\left(\frac{E^2}{m_2c^2}\right)(\eta - 1)}\right)^2. \tag{2.27}$$

In contrast to the nonrelativistic case, the formula (2.17), formula (2.27) possesses in its right part value $\alpha = \frac{E\sqrt{E_1 + m_1c^2}}{m_2c^2}$ which depends on the total energy of electron E_1 the structure of

which includes also kinetic energy *E*k . But dependence of value α on *E*k not strong as the total energy structure includes rather big rest energy of electron *m*1*c*2.Considering that $E_1 = \frac{m_1c^2}{\sqrt{1-\beta^2}}$, where $\beta = \frac{V_1}{c}$ is the relative speed of the photoelectron, we find:

$$\frac{\alpha^2}{2}=\left(\frac{E}{m_2 c}\right)^2\frac{m_1}{2}\left(\frac{1}{\sqrt{1-\beta^2}}+1\right). \tag{2.28}$$

Substituting the equation (2.28) in the equation (2.27) and considering that

$E_k = m_1 c^2\left(\frac{1}{\sqrt{1-\beta^2}}-1\right)$, we get:

$$2c\sqrt{(1-\mu)}=\frac{E}{m_2 c}\left(\cos\theta+\sqrt{\cos^2\theta+\mu\frac{m_2}{m_1}(\eta-1)}\right), \tag{2.29}$$

Where $\mu=\frac{2}{\frac{1}{\sqrt{1-\beta^2}}+1}$.

Considering that $2c\sqrt{(1-\mu)}=\sqrt{\frac{2E_k}{m_1}\Big/\left(1+\frac{E_k}{2m_1c^2}\right)}\approx V_1$, at $1 \gg \frac{E_k}{2m_1c^2}$, we find:

$$V_1\approx\frac{E}{m_2 c}\left(\cos\theta+\sqrt{\cos^2\theta+\mu\frac{m_2}{m_1}(\eta-1)}\right) \tag{2.30}$$

The formula (2.30) allows to consider relativistic effects at the photoeffect, in case of rather big speeds of photoelectrons. Thus, in contrast to (2.17), relativistic coefficient μ is introduced under the root. The calculation of dependence μ(β) shows on fig. 10, relativistic effects while calculating distribution of photoelectrons escape, can be neglected and (2.17) can be used while the photoelectron speeds read approximately half the value of the light speed in the vacuum.

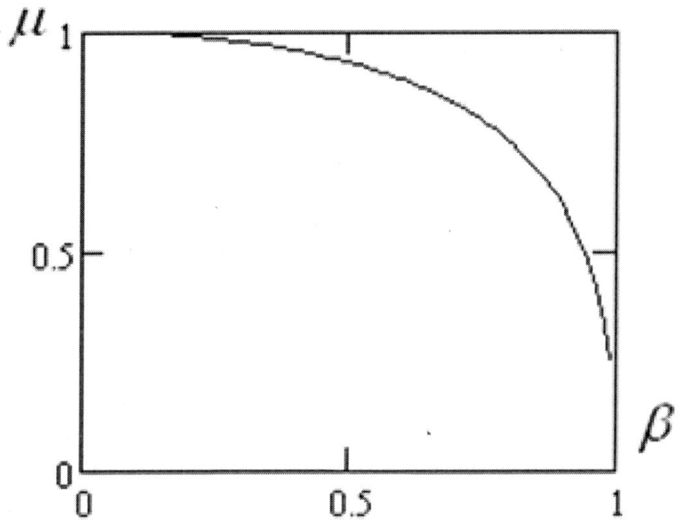

Figure 10. Dependence of relativistic coefficient β on relative speed of photoelectrons $\beta = \frac{V}{c}$

3. CONCLUSION

The laws of formation of the impulse of electromagnetic radiation in dielectric environment for conditions self-induced transparency are considered. The insufficiency of the description of such impulse with the help of the equations Maxwell - Bloch are shown. The impulse of electromagnetic radiation in conditions of a self-induced transparency submits to nonlinear equation of Schrödinger with logarithmic nonlinearity. The way of connection of an average number filling and energy of the impulse taking into account energy saturation of environment are offered. The calculation of a electrical component of the impulse is submitted.

Angular distribution of photoelectrons is investigated during the inner photoemissive effect for two variants: quantum of light basically reveals wave and basically corpuscular properties interacting with orbital electron. Distinction in angular distribution of photoelectrons for these variants is demonstrated. If electromagnetic radiation shows basically quantum properties during a photoeffect there is an emission of photoelectrons on a direction of movement of quantums. It corresponds Einstein's to formula. In Einstein's formula there is no corner of a start of photoelectrons. Angular distribution in the second variant is investigated for the nonrelativistic and relativistic cases.

REFERENCES

1. A. N. Volobuev, V. A. Neganov, 2002 The Electromagnetic Envelope Soliton Propagating in Dielectric. Technical Physics Letters. S-Petersburg. 28 2 15 20 .
2. R. K. Dodd, J. C. Eilbeck, J. D. Gibbon, H. C. Morris, 1982 Solitons and Nonlinear Wave Equations. Harcourt Brace Jovanovich, Publishers. London, New York, Toronto. 545 588, 601.
3. M. J. Ablowitz, H. Segur, 1981 Solitons and the Inverse Scattering Transform. SIAM. Philadelphia. 374 378 .
4. Volobuev A.N. Modeling of Physical Processes, to Describe by Nonlinear Schrodinger Equation. Mathematical modelling. Moscow. 2005 17 2 103 108 .
5. G. L. Jr Lamb, D. W. Mc Laughlin, Aspects of Solitons Physics. In collection under edition of R. K. Bulllough, P.J. Caudrey. Solitones. Springer-Verlag, Berlin-New York, 1980 59 102 .
6. V. B. Berestetskij, E. M. Lifshits, L. P. Pitaevskij, 1989 Quantum electrodynamics, Moscow Science. 192 249.
7. A. S. Davidov, 1963 Quantum mechanics, Moscow Phys . matlit. 335 366.
8. J.P. Birnbaum, 1967 Optical quantum generators. Moscow. Soviet radio, 49 50.

9. G. B. Whitham, 1974 F.R.S. Linear and Nonlinear Waves. A Wiley-Interscience Publncation. John Wiley & Sons, London, New York, Toronto. 575

10. R. Ditchbern, 1965Physical optics Science M. 413 419, 497.

11. N. Ashcroft, N. Mermin, 1979Solid-state physics, Moscow World. 1 82

12. E.L. Nolle, 2007 Tunneling mechanism of the photoeffect in metal nanoparticles activated by caesium and oxygen. Moscow, UFN177 № 10. 1133 1137.

13. A. I. Mihajlov, I. A. Mihajlov, A doubl enu clear photoeffectin relativistic region. Angular and energy distributions of photoelectrons. Moscow, JETP. 1998 114 №. 5. 1537 1554.

14. V. G. Levich, 1962 The Course of theoretical physics. 2 Moscow, PhysMathGiz. 658

15. M. A. Blohin, 1957 Physics of Roentgenrays Moscow, State Publishing House. 261

16. D. A. Steigerwald, W. F. jr Egelhoff, 1988Phys Rev. Lett 60 2558